skydreamers
a saga of air and space

skydreamers

a saga of air and space

ESSAYS BY

Michael Benson

Stephen White

Simon Winchester

AUTRY NATIONAL CENTER

STEPHEN WHITE EDITIONS

ISBN-10: 0-615-37891-6
ISBN-13: 978-0-615-37891-6

Published by
Stephen White Editions
P.O. Box 1664
Studio City, California 91614
 and
Autry National Center
4700 Western Heritage Way
Los Angeles, California 90027-1462
TheAutry.org

This catalogue is produced for the exhibition *Skydreamers: A Saga of Air and Space*,
on view from at the Autry National Center from December 10, 2010 to April 17, 2011,
and to commemorate the 100th anniversary of the first air meet in the United States
at Dominguez Hills, California, in January 1910.

Book design: Michele Perez
Copy editing: Sherri Schottlaender
Copy editing: Marlene Head (Autry National Center)
Scans by Digital Fusion

Front cover: Unknown photographer, *"Answer from God,"* first in a series of articles and
   illustrations published in the Japanese newspaper *Yomiuri Shimbun,* 1957
Back cover: Unknown photographer, *Otto Lilienthal in flight,* circa 1870–90s
Inside cover: NASA, *Hubble Space Telescope, ultra-deep field, constructed from 11 days of
   observations,* 2004 (printed 2010)
Page 2: Unknown photographer, *Otto Lilienthal in flight,* circa 1870–90s
Page 3: Will Connell, *Astronomer looking through the Hale-telescope at Palomar Observatory,
   California,* 1949
Page 4: NASA, *Rendezvous of Gemini 6 and Gemini 7,* 1965
Pages 6-7: Jesse F. Santos (attributed), *Star Wars painting depicting battle,* circa 1980
Page 8: Roy Knabenshue, *Tourists riding in the gondola of Knabenshue's dirigible, St. Louis
   World's Fair,* 1904
Pages 10-12: Keystone View Company, *Roger Hayward seated on an exact scale of the moon
   reduced to 38 feet at Griffith Park Observatory,* 1934

Printed in Spain by Grupo Jomagar

table of contents

An asterisk* beside a caption number indicates there is
additional information available in the checklist

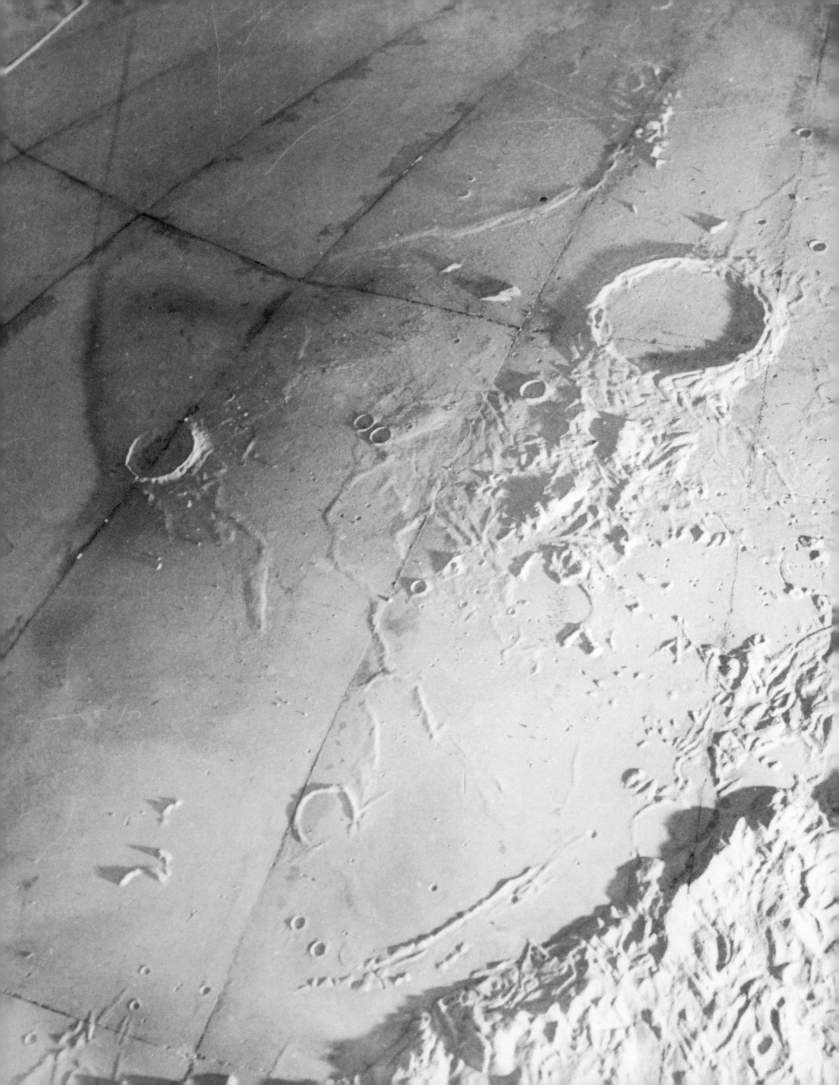

Photographed June. 11. 1901

foreword

Journeys into the unknown inhabit a particularly powerful realm within the West of our imagination. Migrations and voyages of discovery, both individual and collective, are at the heart of Western stories. It seems altogether fitting, then, that *Skydreamers* should find a home at the Autry National Center of the American West.

One hundred years after the first air meet in the United States launched planes into the sky from Los Angeles County's Dominguez Hills in 1910, the West remains at the center of a global quest extending into the sky and out toward the far reaches of the universe. Today the scientists, engineers, and artists who live and work in the West are part of global networks of technology and culture, as indeed they have been since the first Spanish ships appeared off the coast of California some five hundred years ago. The photographs in this catalogue show the West as part of that world, a part that perhaps more than any other embodies the spirit that continues to propel us onward.

The Autry would like to thank Stephen White for initiating this project and for collaborating with our team in such a gracious, professional, and inspired way. We also extend our thanks to contributors Michael Benson and Simon Winchester for their essays, which help us see these photographs, like the American West itself, as part of a bigger picture. The undiscovered country remains vast beyond our wildest imagining, and new adventurers are born every day.

JONATHAN S. SPAULDING
EXECUTIVE DIRECTOR
MUSEUM OF THE AMERICAN WEST, AUTRY NATIONAL CENTER

13
Unknown photographer
Professor Langley's
"Aerodrome" model plane
1901
silver print

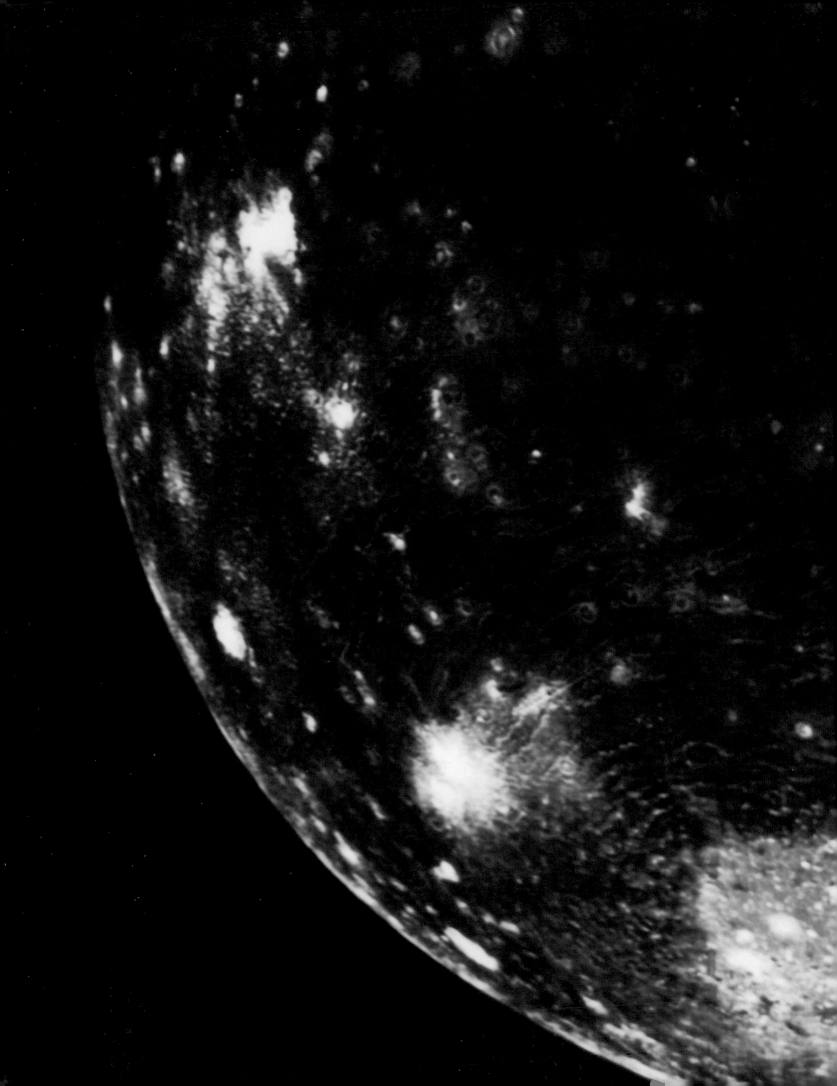

introduction

A Saga of Air and Space

STEPHEN WHITE

I have always been a skydreamer, although my desire to conquer space has only been voyeuristic. When the Autry National Center agreed to host *Skydreamers*—with much of the material gathered from my own collection of aviation and space photographs—I envisioned a catalogue and exhibition that would trace the beginnings of flight through the aviation period and into an exploration of the greater universe. I wanted to make people conscious of the aesthetic qualities of flight and space exploration images as well as their informational content. As the selection process evolved, however, and the sections began to sort themselves out through the works to be shown, more complex issues evolved. These issues go far beyond the original intent of the exhibition.

Skydreamers, both the exhibition and catalogue, has morphed like the expanding universe, moving beyond my aim of showing the development of flight and space exploration through photographs. Industrialization and technology changed our world dramatically, and in an astonishingly brief time period. One of those evolving technologies, photography, gave us a method to capture and study our world and what exists in the heavens beyond. But photography does not answer the bigger question: Why?

Of course, the quest to conquer and understand space long preceded the Industrial Age. The first aerial success was an eighteenth-century balloon ascent, and the progression from that first balloon flight has led to discoveries that have fascinated and occupied mankind for the past two hundred years. Although it took us 120 years to move from that free-floating balloon ascent in 1783 to the first heavier-than-air flight in 1903, it only took another 66 years from that first 12-second flight to place men on the moon. Twenty-one years later, in 1990, the Hubble Telescope was launched into space. After numerous repairs and adjustments, the Hubble began to read further into the universe, and its eye allowed us to explore and capture, through photography and film, regions of the universe previously undetected (and consequently, until their recent detection, beyond our realm of thought). As I read about the saga of those pioneers who first sought to understand flight, those who worked to conquer the dynamics of controlled flight, and the scientists examining and exploring the farthest reaches of space, I found myself revising the concept of the catalogue and exhibition. My intent had always been to show the beauty of these ideas and illustrate how photography was the ideal vehicle to convey the shapes, forms, and spatial relationships that aviation and space exploration produced. In my attempt to make decisions about content based on aesthetics as well as history, I began to see the world differently than I had previously envisioned it.

183
NASA
Mosaic of Jupiter moon Callisto, taken from Voyager 1 (detail)
1979
chromogenic print

Although I have always preferred scientific explanations to religious ones, I now more clearly understand how our world is merely a microcosm that exists within a microcosm—our solar system—which is itself a microcosm within the width and length of our Milky Way galaxy bursting with its billions of stars. Astronomers, those Darwins of outer space, have opened my eyes. They bring insight into the universal order and disorder, providing a clear perspective on issues that surround the universe and our insignificant role within it. It is challenging to speculate about whether their findings in this modern information age have begun to impact traditional ideas about the relationship between God and humankind, ideas that philosophers and religious leaders have explored since the beginning of civilization. Just as Darwin upset the evolutionary apple cart for creationists, astronomers are discovering new information that continues to reduce our role and significance in a universe much broader, wider, deeper, and more complex than anything we could have imagined. Despite this, religion remains a dominant force around the world, and the ideas, interpretations, and fantasies of past religious leaders still dominate popular thought, while new religious leaders are as common as asteroids.

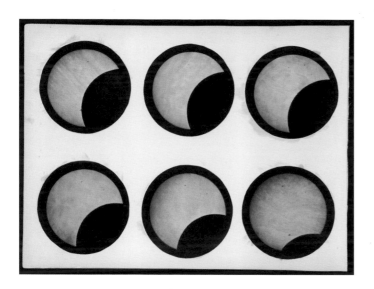

One can only wonder. Who am I? Why am I here? What purpose does a life have, or for that matter, is there any reason for the Earth itself to exist? These questions, asked for millennia, now can be examined in a broader context. People around the world have always looked to the sky to seek answers, treating natural phenomena as supernatural prior to the advent of science. Today, astronomers raise provocative new questions about our planet and the broader universe as they read and interpret information for us based on what they have discovered via the Hubble and other space telescopes. As they decipher the data provided by the telescopes, they are introducing us to new complexities in the universe, including events that occurred billions of years ago in deep space far beyond the power of any earthly vision. These discoveries, made on almost a daily basis, can help us understand the why and how of our existence, perhaps far better than can any minister, priest, rabbi, or mullah.

The title of the photograph on the cover of this catalogue, *An Answer from God*, suggests that by granting us access to technology, a higher power controls our ability to move into space beyond the limits of our own planet's gravitational pull—it is as if a supreme being has allowed us to step out of our iron shoes and float gently through the universe. But this god (whatever its form) who answered our prayers had some surprises in store for us once we focused our eyes and minds on the distant reaches of the universe: galaxies new and unknown surfaced like so many grains of sand on endless beaches, expanding exponentially beyond the limits that could be absorbed by any mind. Space itself appeared endless, stretching as far as the concept *pi* stretches in mathematics, a dark and light expanse, contracting and expanding. In this newly uncovered universe, black holes swallowed galaxies while stars were born as frequently as children are born to mothers.

In recent years the word *billions* has become commonly used when discussing economics, but it almost seems inadequate when applied to the number of stars. Our existence is centered around only one star (the sun) in the Milky Way galaxy, which alone contains around two hundred billion stars—and the Milky Way is only one of more than one hundred billion galaxies large and small in the known universe. If there is no way of knowing how many additional galaxies exist beyond our visual reach, then where exactly are we, and is our existence any more than some glorious astronomical accident?

These questions have led me to a metaphor I can comprehend. They help me to understand both the earth's relationship to the universe and my own relevance to the earth. I imagine the earth as a grain of sand and the universe as an ocean so vast its shores are ultimately unknown. And each of us, all six billion plus, is a grain of sand alongside the vast ocean that is our earth.

We humans have often wondered about the existence of other forms of life. Are we alone, or have circumstances allowed other planets to form in a way that might sustain

170°
Lick Observatory
Orion Nebula (detail)
circa 1950
(printed by R. P. Peterson)
silver print

life as ours has? In an effort to find out, NASA's Kepler Mission has launched a telescope to study a portion of the Milky Way in a search for other planets in our galaxy near in size to our own. Their search has revealed hundreds of other planets around distant stars.

This newly altered and constantly evolving vision of our universe forces us to ask questions: Is our God really a benevolent heavenly father? Or perhaps the time has come to view God as nothing knowable, but as the forces of nature that open expanding stretches of the universe and bring together new galaxies—each with billions of stars—much as we on Earth build new cities. Hermes Trismegistus, a Greek-Egyptian deity, is credited with the saying, "As above, so below," and his followers attribute occult mysteries to this concept. In fact, there are many parallels here on Earth to what transpires in the universe.

Space is full of contradictions: there is both chaos and order throughout the universe, just as there is in our day-to-day life. Since ancient times, mystical works have offered insights that can lead to a deeper understanding of the universe and universal ideas. However, unlike mystics or visionaries who reflect and gaze inward, astronomers have been studying the universe through telescopes for four hundred years, first through the primitive lenses of Galileo, Kepler, and Tycho Brahe's telescopes, and eventually through a series of ever more powerful telescopes and lenses. Earthbound astronomers began to unlock the secrets of the universe in a way that no philosopher had been able to do. Spurred on by their great curiosity and thirst to understand the dynamics that produced the orbiting ball they inhabit, they have used the telescope as a method of capturing secrets from the universe. The more sophisticated their technology, the more expansive the universe they have encountered. Astronomy confronts us with the ultimate reality: the endlessness of outer space.

NASA's Marshall Space Flight Center History Office website informs us that as early as 1923 a Romanian-German visionary named Hermann Oberth suggested that a telescope could be propelled into space using a rocket. According to Oberth, "If there is a small rocket on top of a big one, and if the big one is jettisoned and the small one is ignited, then their speeds are added." The work of Oberth—considered to be one of the fathers of modern rockets along with Robert Goddard of the United States and Konstantin Tsiolkovsky of the Soviet Union—helped make the science of rocketry a reality.

Technology allowed our civilization to advance more rapidly during the last century than at any time in the past. When we launched the Hubble, the universe became visible to us, but the seed for that launch was planted much earlier, during a time when Benjamin Franklin flirted with ladies of the French court and Americans were still celebrating their newborn independence: it began when that that first unoccupied balloon lifted over the Champs des Mars, witnessed by countless thousands.

That arc—from the first successful encounter with the air to the technological triumphs that led to space exploration and the discovery of distant galaxies—with all its profound influence on artists and the public alike, moved and inspired me in the process of organizing *Skydreamers*, and I hope it will do the same for you.

PART I: FLIGHT

By permission of the Patentees, Messrs Henson, Colombine, and

THIS ENGRAVING of the FIRST CARRIAGE, the "ARIE

is respectfully inscribed, to the Directors of

THE AERIAL TRANSIT COMPAN

FIRST UPWARD, THEN ONWARD

SIMON WINCHESTER

The emotion is certainly as old as humankind. Probably it was born even earlier, the first time the first hominid creature glanced upward at the sky or down across a chasm and felt a curious pang of an unrecognized feeling: envy. Although some incomprehensible something kept him rooted to the rocks and savannah where his feet were planted, there were creatures nearby, quite unlike him in form and size, which were somehow not so mysteriously bound to the earth. At any moment of their choosing, they could take off into the air and *fly*.

Even before people developed the concepts involved with this curious animal skill, they were observing it in action and searching for the words to describe it. The first human lexicons, whether they were constructed by the peoples who lived in Ur or Crete, Cairo, Machu Picchu, or Chang'An, are all enriched from the very earliest times with words that define the insects and birds that could perform this gravity-defying miracle. Moreover, in all of the tongues spoken by the civilizations noted above—and especially so in the English that was amalgamated from them—the order of the linguistic recognition was always exactly the same. First there came a word for the beast; next there appeared a word that defined the peculiarity of its ability to defy gravity; and finally there was a name given to the appendage that seemingly allowed it to do so. It is a *bird*, the language says. It *flies*, it continues. And it does so, even if at first inexplicably, by the employment of something that is to be called a *wing*.

Bird—fly—wing. It was always the same trinity of ideas, and it was always a trinity inserted into the language in exactly the same historical order. Insect—fly—wing. Bee—fly—wing. The *Oxford English Dictionary* confirms the sequence: it shows us, for example, that the first use of the word *bird* (which was initially, and very oddly to our ears, spelled as *brid*) occurred some time around AD 800. The concept of *flying*—the action word upon which this book and exhibition are essentially predicated—swept into the lexicon almost exactly two centuries later. Finally, a further 175 years on, came the first appearance in the literature of the word *wing*, the mechanical device that apparently allows the *brid* to accomplish its *flying* in the first place.

Thus it was, over a period of almost five medieval centuries, that English-speaking humankind first conceived and defined the mechanism behind this universally envied marvel, flight. In AD 800 we knew and named the creatures that could do it. By AD 1275 we knew more or less *how* they did it. It would take almost as many years more, years of envying and dreaming and wondering, before Westerners would come up with means of accomplishing the marvel themselves.

2
Constance Warburton
The First Carriage Ariel
1843
watercolor copy of engraving

had spent so many months making, and he managed to keep it ten feet up in the air for twelve long seconds, during which it flew, fully under his power and control, across one hundred and twenty unforgotten feet of sand in the Outer Banks of North Carolina.

A photograph was duly taken, allowing this feat to still be accepted today by most as the first true controlled flight in a heavier-than-air machine. The Wright Brothers remain unforgotten, and the craft they engineered eventually became a prize possession of the Smithsonian's National Air and Space Museum: the Wright Flyer, Number 1.

The dream had at last been realized, the envy finally assuaged. Now flying became a skill to be put to practical use, and we were suddenly able to see the world (and move across it) as never before. Crossing from dune to dune on the Outer Banks—so painstaking of contrivance and so difficult of achievement—soon became no more than a trivial matter. The necessary mechanical and technical specifications for flight evolved quickly to a point where passing from country to country in the air soon became a commonplace event, after which came the steady development of machinery that would enable fliers to pass across bodies of water, from island to island, from continent to island, and in time—the most signal achievement of all— to move from continent to continent over vast bodies of water.

Newspapers offered prizes to those who could do so first. But great seas are not kindly entities over which to fly: if your aircraft somehow fails, where, exactly, do you put down? No pilot leaves the chocks for a transoceanic flight even today without remembering the first axiom always offered in flight school: *take-off is voluntary, but landing is compulsory.* And in the middle of an ocean it is self-evidently true that not only is there nowhere to land, *there just is no land. No land at all.*

Those who pioneered the practice of flying over seawater knew this all too well. Crossing a large expanse of sea perhaps didn't trouble Louis Blériot much when he flew his tiny monoplane across the English Channel from Calais to Dover in 1909, just six years after the events at Kitty Hawk. Although Blériot boasted about being alone above "an immense body of water" for fully ten minutes (even though he was only 250 feet up in the air), he also had the comfort of knowing that a French destroyer below him was monitoring his flight, ready to save him if he ditched. During most of his 37-minute crossing he could see the coast of France behind him, and by peering ahead, the white cliffs of England before him. Blériot won the £1,000 that Lord Northcliffe had offered through his newspaper, the *Daily Mail*, and he became—not least because of his boulevardier's

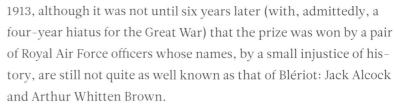

moustaches and his reputation as a barnstorming air racer—an immediate superstar and very much the ladies' heartthrob.

The Atlantic Ocean was the supreme challenge, however, and Lord Northcliffe put up ten times his initial sum for anyone who dared try it. He announced his challenge in 1913, although it was not until six years later (with, admittedly, a four-year hiatus for the Great War) that the prize was won by a pair of Royal Air Force officers whose names, by a small injustice of history, are still not quite as well known as that of Blériot: Jack Alcock and Arthur Whitten Brown.

The venture was Alcock's idea, conceived while he was imprisoned by the Turks after ditching his fighter plane in the sea near Gallipoli. "Why not have a bash?" he asked. In the summer of 1919, the pair dismantled their plane—a stripped-down long-range Vickers Vimy biplane, its bomb bays to be filled with extra fuel—and crated it up so it could be sent by ship to Newfoundland. There they built their own runway for takeoff. They did not know where they might land: it could be a field or a beach or a country lane (it turned out to be an Irish bog).

Plenty of others were trying for the same prize, among them an American, Albert Cushing Read, who flew a seaplane to the Azores, stayed there for a week, and then flew on to Portugal. His trip took eleven days, and American warships were stationed under his path every fifty miles along the proposed route. However, because Lord Northcliffe had decreed that his prize was for a nonstop journey achieved in fewer than 72 hours, Mr. Read did not win it, nor did an Australian tearaway named Harry Hawker, who tried it in a Sopwith Atlantic (an experimental long-range plane). When the craft's engine overheated, Hawker spotted an eastbound ship five hundred miles short of Ireland and ditched. He was picked up and went home by sea, but because the ship had not yet acquired a radio, its crew could not tell anyone of his rescue—instead, Mr. and Mrs. Hawker were shocked to get an official black-bordered telegram from King George offering his condolence for the supposed loss of their son.

The dashing British aviators—Jack Alcock in a blue serge suit and Arthur Brown in his Royal Flying Corps uniform—set off on the morning of Saturday, June 14, with 865 gallons of fuel and a pair of small black cats named Twinkletoes and Lucky Jim. They had horrendous problems: up at 12,000 feet their instruments froze solid, their radio broke, and their exhaust pipe ruptured; Brown actually had to climb onto the wings to break off ice. They became disoriented trying to watch the stately heeling of the stars in order to navigate and went into a spin down through the clouds, almost hitting the waves. When they finally arrived over the coast of Ireland, they could not find a place sufficiently free of rocks on which to land, at last spotting the masts of a radio station. They circled it

FLIGHT FANTASIES

Ever since the first balloon launch, the public h

fascinated with the idea of overcoming gravity

ing freely, like birds. The widely viewed, hand-

engraving of the Montgolfier brothers' experim

balloon ascending over Paris marks the beginn

people's fascination with aerial flight, and with

period of time afterward, balloon shapes could

everywhere in eighteenth-century France in t

hats and dresses, pieces of furniture, and knic

During the nineteenth century many cont

fantasize about conquering the air. Seventy-fi

Otto Lilienthal

Otto Lilienthal, am 22. 5. 1848 geboren, handwerklich besonders begabt und interessiert, ging mit 14 Jahren auf die Gewerbeschule über, wurde Mechaniker, Konstruktionszeichner und Ingenieur und machte sich dank eigener, nützlicher Erfindungen und Konstruktionen bald selbständig. Sein Bruder Gustav wurde Bauarchitekt. Beider eigentliche Liebe aber gehörte der Fliegerei – die es im heutigen Sinne damals noch garnicht gab. Schon als Student der Berliner Gewerbeakademie begann Otto Lilienthal mit Messungen und Berechnungen des Winddrucks und der Tragfähigkeit der Luft. Nach Rückkehr aus dem Kriege 1870/71, den er als Einjährig-Freiwilliger beim Garde-Füsilier-Rgt. mitmachte, setzte er gemeinsam mit dem Bruder seine Versuche fort, den Menschen das Fliegen zu lehren.

10a,g

UNKNOWN PHOTOGRAPHER ***Otto Lilienthal in flight***
circa 1870–90s (printed circa 1920)
silver prints

12

ROL AGENCY ***Octave Chanute gliding***
circa 1904
silver print

14

HARRY ELLIS   *10 photographs depicting balloon and ascent of*
Count Charles de Lambert and Mrs. Lambert
circa 1905
silver prints

John Daniels
*Orville Wright making the first
flight, 120 feet in 12 seconds*
1903
silver print, enhanced

HISTORY OF FLIGHT

Controlled flight began with Orville Wright's twelve-second flight in a piloted, powered, heavier-than-air ship at Kitty Hawk, North Carolina. From that moment forward, the race was on to develop more powerful airplanes and make them commercially viable, the end prize being the first to achieve a major and consistent conquest of the air.

Wilbur Wright traveled to France in 1908 and shocked the French with his ability to fly his plane in circles. The following year Louis Blériot became the first to fly over the English Channel, and that same year the first recognized air meet was held in France. In 1911 Cal Rodgers became the first American to fly cross-country—though he had more than thirty crashes during his route from one side of the country to the other—and Eugene Ely landed his craft on the deck of the USS *Pennsylvania*, signaling the beginning of naval aviation.

In the years to come, many records would be set and many heroes anointed, even as the death toll climbed due to the challenge and danger of flight. Aviation heroes from Wilbur Wright to Charles Lindbergh tested the limits of powered flight. Others, such as Lincoln Beachey, Bessie Coleman, Antoine de Saint-Exupéry, and Amelia Earhart, gained recognition for their exploits and risk taking, and each met an early death. As early as 1912, as the science behind aviation was still developing, Count Ferdinand von Zeppelin devised a method for carrying passengers in his industrious invention, and by the time Lindbergh landed in Paris in 1927, airlines such as Air France, Lufthansa, and KLM had begun to set up a network of flights within Europe. Even as great advancements continued to be made in general aviation, the thinking and focus of engineers and the public in general turned to the race for space.

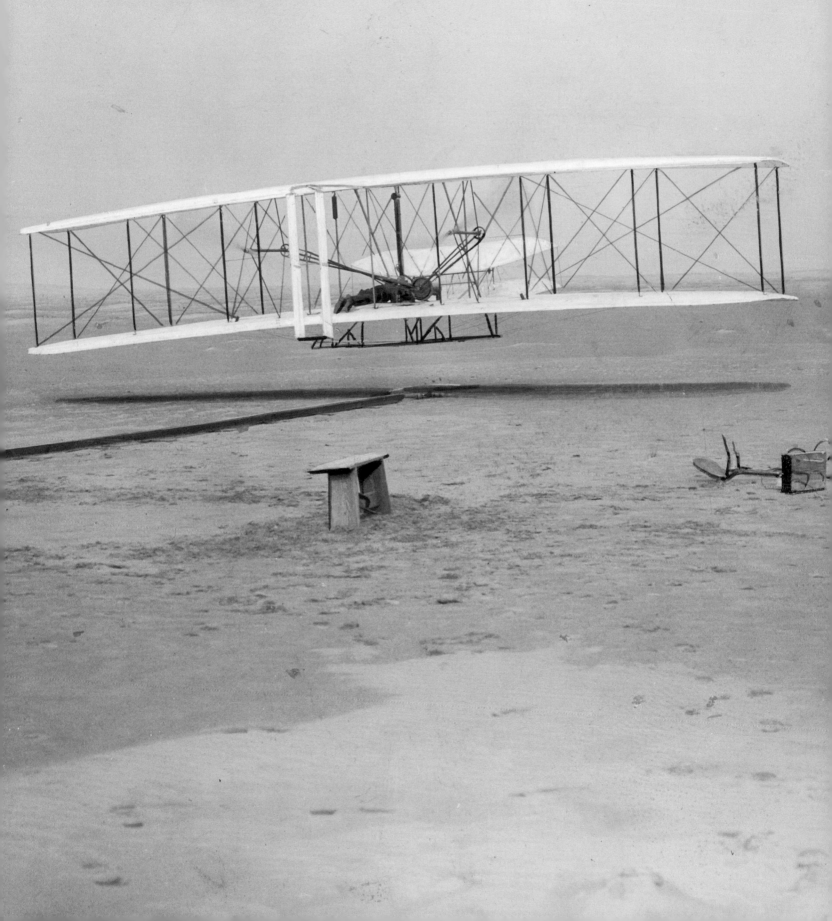

GEORGE GRANTHAM BAIN　　*Wilbur Wright flies around the Statue of Liberty*
1909
silver print

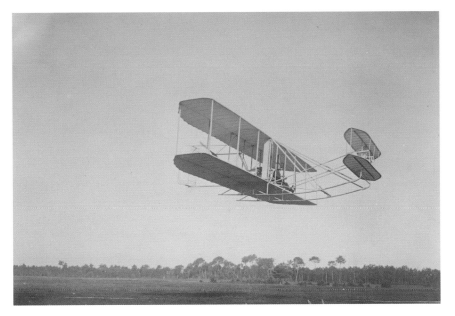

17a

ROL AGENCY   *Wilbur Wright taking off at Le Mans, France*
1908
toned silver print

17b

ROL AGENCY   *Wilbur Wright in the air at Le Mans, France*
1908
toned silver print

73. Santos Dumonts Hydroplane. Courbevoie
France

18

HARRY ELLIS *Santos Dumont's hydroplane at Courbevoie, France*
1906
silver print

19

BROWN BROTHERS *Glenn Curtiss in his June Bug*
1908
toned silver print

20

ROL AGENCY   *Louis Blériot nearing the English coast on the first*
              flight across the English Channel
              1909
              silver print

24*

UNKNOWN PHOTOGRAPHER   *Eugene Ely landing his plane on the deck of the*
                        USS Pennsylvania
                        1911 (printed circa 1920)
                        toned bromide enlarged print

21

APEDA STUDIO   *First water rescue, by Hugh Robinson, Lake Michigan*
                1911
                silver print

UNKNOWN PHOTOGRAPHER   *An early morning flight, possibly by a World War I*
                        pilot in his Morane–Saulnier
                        circa 1915
                        silver print, inscribed

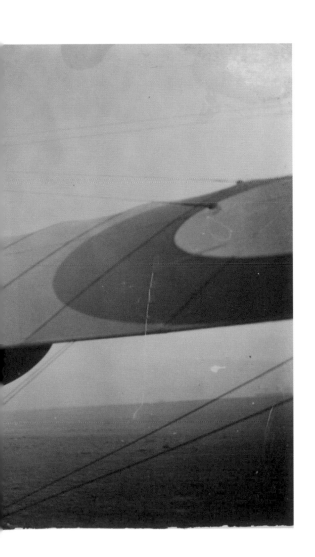

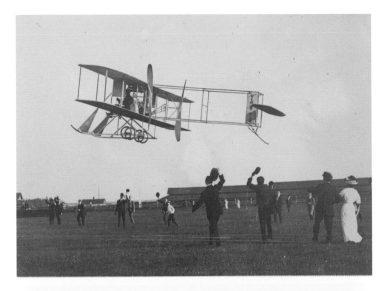

23a

UNKNOWN PHOTOGRAPHER *The* Vin Fiz *takes off from Sheepshead Bay, Long Island,*
on first cross-country flight
1911
silver print

23b*

UNKNOWN PHOTOGRAPHER *The* Vin Fiz *crashes in Compton, California*
1911
silver print

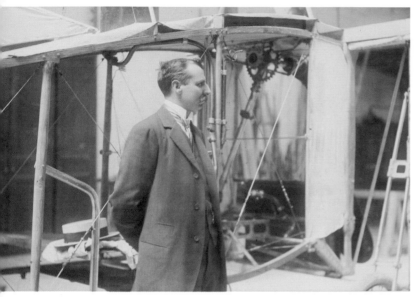

27

HENRI MANUEL   *Portrait of Louis Blériot given by the pilot to his physician,*
Dr. Aric, after a crash in Istanbul
1909
silver print, dedicated on mount

26 *

ROL AGENCY   *Wilbur Wright being photographed at Camp d'Auvours,*
Le Mans, France
1908
silver print

32*

UNKNOWN PHOTOGRAPHER

*Lincoln Beachey ready to test his Taube aircraft
at Ocean Beach in San Francisco*
1915
silver print

28*

WILLIAM PRESTON MAYFIELD

*Arthur L. Welsh, the first American Jewish
aviator, sitting in a Wright flyer*
circa 1910
silver print

31*

UNKNOWN PHOTOGRAPHER *Portrait of Amelia Earhart in her plane*
circa 1937
silver print

34*

JEAN THEVENET *Antoine de Saint-Exupéry departing*
(attributed) *Gare du Midi, Brussels*
circa 1940 (printed circa 1950)
silver print

33*

NASA, UNKNOWN PHOTOGRAPHER   *Bessie Coleman, the first African American woman aviator*
circa 1921
digital black-and-white print

35*

BOB SEIDEMANN   *Chuck Yeager, test pilot*
1990s
silver print

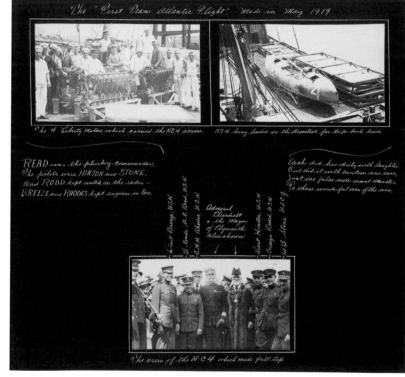

UNKNOWN PHOTOGRAPHER   *Various snapshots depicting the first
transatlantic flight*
1919
silver prints

The "First Trans-Atlantic Flight". Made in 1919

At Trepassy Bay, New foundland.

The "First Trans-Atlantic Flight". Made in May 1919

NC-1   NC-3   NC 4

Taking off for the Azores on May 16. 1919, from Trepassy Bay.

NOTE
NC-1
&
NC-3
were
wrecked
at
Azore
Islands

NC 4 then
proceeded to
Lisbon Portugal

NC 4
Arriving at Plymouth, England
May 31, 1919

NC 4
Squataim off at Plymouth, England

NC 4
at Plymouth, England

NC 4
being dismantled for return to U.S.

38

HORACE KEARNEY   *Rockwell Field planes in flight over San Diego*
1918
silver print

42

UNKNOWN PHOTOGRAPHER *Paris Air Show*
1934
bromide print with
watercolor enhancement

40

ASSOCIATED PRESS (AP) Hindenburg *over New York City moments before it
crashed and burned at Lakehurst, New Jersey*
1937
silver print with letterpress

44*

UNKNOWN PHOTOGRAPHER   *Harold Osborne's parachute snagged on the tail of a plane*
1931
silver print

45*

ARCHIE HALL   *A page in his scrapbook showing travel by commercial
planes in Europe*
1927
silver prints

ON OUR WAY, THE NUMBER OF THE PLANE AND THE DESTINATION. EUROPEAN PLANES IN COMMERCIAL SERVICE ARE CONSTANTLY IN TOUCH WITH THEIR HOME AERODROMES BY RADIO. ALL PLANES CARRY A SENDING AND RECEIVING SET HAVING A RANGE OF ABOUT 200 MILES AND ALL ARE TELEPHONE SETS. WEATHER REPORTS ARE GIVEN DURING THE FLIGHT BY THE AERODROME. BACK IN THE FLYING CONTROL ROOM, THE COURSE OF EACH PLANE IS PLOTTED AND MARKED ON THE MAP AFTER EACH REPORT IS GIVEN.

WAS REPAIRING THE VALVE ROCKER ARM, WE HEARD ANOTHER PLANE. DON'T KNOW WHERE IT CAME FROM, BUT IT ALSO WAS AN AIR UNION PASSENGER SHIP. BEFORE WE STARTED TO DESCEND, THE RADIO MAN HAD TOLD THE AERODROME ABOUT OUR ENGINE TROUBLE. AND THEY HAD SENT ANOTHER PLANE FOR US. BY THE TIME IT ARRIVED, THE RADIO MAN, WHO DOUBLED AS THE MECHANIC, HAD FIXED THE ENGINE. HE STEPPED AWAY FROM THE GANG AND SIGNALLED THE OTHER PLANE ALL WAS WELL ONCE MORE. THE RELIEF PLANE PILOT

GETTING BACK TO OUR PLANE, FOUR PASSENGERS SAT IN NICE EASY CHAIRS IN THE COMPARTMENT IN FRONT OF THE PILOT'S ROOM. EIGHT OTHERS COULD SET IN THE SOFT CHAIRS TO THE REAR OF THE PILOT'S ROOM. THE PILOT SAT SO HIS HEAD WAS STICKING ABOVE THE TOP WING. AT THE REAR OF THE PASSENGER COMPARTMENT WAS A TOILET COMPARTMENT, WITH THE BAGGAGE ROOM IN BACK OF IT.

ABOUT TWO HOURS AFTER WE HOPPED OFF, I SAW THE RADIO MAN WHEELING IN HIS AERIAL WIRE. THEN THE PLANE STARTED TO DESCEND. AND CAME DOWN IN A FIELD AMONG A BUNCH OF SHEEP. THERE ARE THE SHEEP AND A PART OF A PLANE WING SHOWN RIGHT ABOVE THIS. THE RADIO MAN, WHO SPOKE SOME ENGLISH, EXPLAINED ONE OF THE VALVES ON THE RIGHT ENGINE HAD "FROZEN." IT HAD REFUSED TO WORK, SO WE HAD TO LAND. WHILE HE

DIPPED IN ANSWER AND SOARED AWAY. A FEW MINUTES LATER WE HOPPED OFF AND FLEW ABOUT TWO HOURS MORE. BUCKING A HEADWIND ALL THE WAY. WHICH CUT OUR SPEED DOWN TO 60 MILES AN HOUR. AND MADE THE RIDING A LITTLE ROUGH. AT 4.45 P.M., WE CAME DOWN FOR THE SECOND TIME. SAME KIND OF TROUBLE, ONLY ON OTHER ENGINE. THE RADIO MAN TRIED TO FIX IT IN THE AIR, BUT COULDN'T. THIS TIME WE MADE A PERFECT THREE-POINT LANDING IN A WHEAT FIELD. NEVER SAW SO MANY PEASANTS GATHER IN ONE SPOT IN SUCH A SHORT TIME. INSIDE OF 15 MINUTES THERE WERE 200 PEOPLE, OLD FOLKS, KIDS AND IN-BETWEENS GATHERED AROUND. PICTURES AT LOWER LEFT AND UPPER RIGHT SHOW PART OF THEM. THE SHIPWRECKED PASSENGERS ARE SHOWN IN FRONT OF THE PLANE AT LOWER RIGHT. ALL KINDS OF PEOPLE THERE WATCHING THE EVENT. GOOD LOOKERS AND SOME NOT SO KEEN. GUESS IT WAS THE FIRST TIME TWELVE PERSONS DROPPED IN SO SUDDEN AND UNEXPECTED LIKE ON THEM. MECHANIC GOT THE VALVE

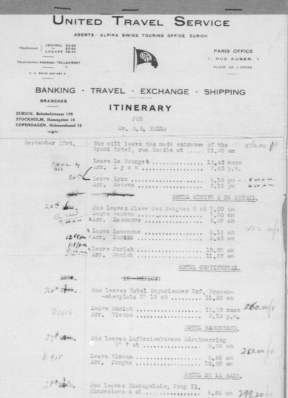

L. VILLMONT

UNITED TRAVEL SERVICE

AGENTS: ALPINA SWISS TOURING OFFICE, ZURICH

TELEPHONE | CENTRAL 33-49 / 33-50 / LOUVRE 56-34

TELEGRAPHIC ADDRESS: TELLGARRET

PARIS OFFICE
4, RUE AUBER, 4
PLACE DE L'OPERA

BANKING - TRAVEL - EXCHANGE - SHIPPING

BRANCHES
ZURICH, Bahnhofstrasse 108
STOCKHOLM, Hamngatan 14
COPENHAGEN, Holmenskanal 15

ITINERARY
FOR
Mr. A. S. MALL:

September 23rd.	Bus will leave the main entrance of the Grand Hotel, rue Scribe at ..... 11.00 am	550.00
	Leave Le Bourget .... 12.45 noon	
	Arr. Lyon .... 2.45 p.m.	
	Leave Lyon .... 4.15 pm	
	Arr. Geneva .... 6.15 pm	
	HOTEL GENEVE & DU BRESIL.	
	Bus leaves Place des Bergues 3 at 7.00 am	
	Leave Geneva .... 7.30 am	
	Arr. Lausanne .... 8.00 am	
	Leave Lausanne .... 8.15 am	
	Arr. Zurich .... 9.45 am	
	Leave Zurich .... 10.00 am	
	Arr. Munich .... 12.55 pm	
	HOTEL CONTINENTAL.	
	IN MUNICH:	
26th am.	Bus leaves Hotel Bayrischer Hof, Promenadenplatz N° 19 at .... 11.50 am	
	Leave Munich .... 12.35 noon	360.00
	Arr. Vienna .... 3.10 p.m.	
	HOTEL RASNERAND.	
27th am.	Bus leaves Luftreisebureau Kärntnerring N° 7 at .... 8.00 am	
	Leave Vienna .... 8.55 am	251.00
	Arr. Prague .... 10.55 am	
	HOTEL DE LA GARE.	
28th am.	Bus leaves Skodapalais, Prag II, Charvatova 4 at .... 9.50 am	249.30
	Leave Prague .... 11.10 am	
	Arr. Dresden .... 12.30 noon	
	Leave Dresden .... 12.45 noon	

46

MARGARET BOURKE-WHITE   *TWA Douglas Luxury Skyliner*
circa 1940
silver print

47

UNKNOWN PHOTOGRAPHER   *Loading cargo into the hold of an American Airlines
Douglas DC-3 at Los Angeles*
circa 1940
silver print

FLIGHT AS ART

Prior to the advent of photography in the first decades of the nineteenth century, an image could be created only through the traditional mediums of art. Photography, announced in 1839, needed about fifty years to get up to speed. The motion photographs taken by Eadweard Muybridge and Étienne-Jules Marey in the late nineteenth century paved the way for artists to begin to see the world in a new way, in stop motion. By the twentieth century, however, even amateurs could capture motion with their cameras. Suprematism, cubism, constructivism, and other modernist painting movements reinvented shapes and forms utilizing geometric angles and constructs, ideas that were also reflected in airplane designs (as well as the machines in flight). When a great artist such as Alfred Stieglitz photo-graphed a flying plane, he produced an aesthetically pleasing but not necessarily unique image. However, Russian artists Kasimir Malevich and Alexander Rodchenko began to see the airplane—and the perspective of the airplane—as opening doors to new directions for art. The arrival of Albert Guyot in Moscow in 1909 led to a fascination with flight that attracted the avant-garde Russian poet Vasily Kamensky, who temporarily abandoned literature for the poetry of flight.

Soon photographers were applying an artistic eye to space and the strange-looking, heavier-than-air bodies that began to fill that space. Later, some pilots emerged as aerial photographers who specialized in capturing flight in an artistic way, as shown in the photographs of Alfred Buckham, Margaret Bourke-White, and more recently, Bob Seidemann, who has produced a large, stunning portfolio titled *Aviation as Art.*

56

EMMANUEL EVZERIKHIN   *Parachutes at Tushino Airfield, Moscow*
circa 1950s
silver print

49*

ALEXANDER RODCHENKO   *Copied montage including an airplane*
1923 (printed later)
silver print

ROBERT DOISNEAU   *L'Aeroplane de Papa*
1934 (printed later)
silver print

53*

UNKNOWN PHOTOGRAPHER   *Albert Guyot in Moscow, Russia*
1909
silver print

55

B. N. YUROV   *Design for a helicopter, Russia*
1909
reverse negative print

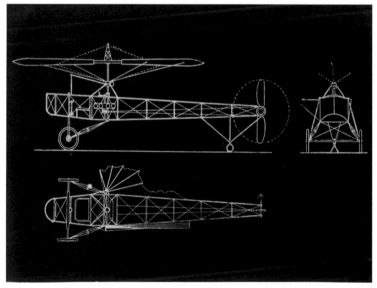

54°

ROMAN KARMEN   ***Aerial photograph***
circa 1932
silver print

61

ROBERT HOYT   *Pan American Clipper at sunrise over Cape Florida
and the old Spanish Lighthouse*
1935
toned silver print

57*

FRANTIŠEK DRTIKOL   *The artist Václav Rabas in an aviator's headgear*
1926
bromide print

58*

AARON SISKIND   *Toy plane as abstraction on a landing field*
circa 1946
silver print

59

UNKNOWN PHOTOGRAPHER   *Subsonic flow on a Delta wing*
circa 1950s
silver print

65*

ALFRED G. BUCKHAM   *Flying over Perth, Australia*
circa 1930
silver print

64

ROBERT A. BURROWS   *Sky patrol*
circa 1939
toned silver print

62*

BUREAU OF AERONAUTICS   *Echelon formation*
1927
silver print

63

MAURICE BRANGER   *Old Ways and New Ways*
circa 1920
enhanced silver print on watercolor paper

66

JAMES DOOLITTLE   *Night flight, Los Angeles*
circa 1938
silver print

68

BOB SEIDEMANN   *The B–2 (Stealth) bomber
photographed from below*
from the series *Aviation as Art*
circa 1990
silver print

69a

*Original notes by Wright Morris written on paper
on the back of the frame for image 69b*
circa 1947

69b*

UNIDENTIFIED FARM SECURITY
ADMINISTRATION (FSA)
AERIAL PHOTOGRAPHER

*A section of Nebraska identified by Wright Morris
as the location for his second photographically
illustrated novel,* The Home Place
circa 1930s
silver print

MARGARET BOURKE-WHITE   *Smoke Screening, New York*
(attributed)   circa 1935
silver print

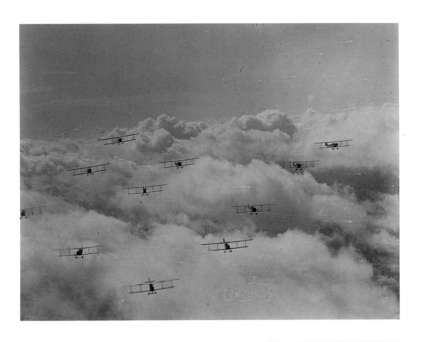

70a–b

WILLIAM PRESTON MAYFIELD   *Two photos of planes in formation flights*
circa 1930
silver prints

73b
Roy Knabenshue
Charles Miscarol in a Blériot
plane in front of stands
1910
silver print

FLIGHT AND POPULAR CULTURE

There is no doubt that flight, beginning with that first Montgolfier balloon ascent, had an enormous impact on the popular mind. Once aviation became more commonplace, the air show was used to focus the public's attention on the competitions they produced. The first air show in America, held at Dominguez Hills, California, was spread over ten days—by its end, the total attendance number had increased to equal half the population of Los Angeles at the time. Air shows multiplied, and with them, thousands of skydreamers sought ways to become masters of the skies. In barnstorming, daring pilots performed stunts individually or in "flying circuses" at county fairs. Heroes and their planes became subjects to be marketed in merchandise, in photo booths shaped like airplanes or balloons, in designs for both clothing and furnishings, and especially in the films that fueled the dreams of many young boys and girls: every boy wanted to be Lindbergh and every girl sought to emulate Amelia Earhart. Books inflamed the imagination of young people, and every aviation success produced multitudes of fans wanting to share the eternal moment. The airplane prevailed everywhere, and within the course of a couple of generations flying became a commonplace reality, one that could be shared by the public at large as passengers in a gigantic aircraft capable of circumnavigating the globe.

76
Fred Wagner
Aerial view of aero meet at
Grant Park, Chicago (detail)
August 1911
ferrotyped silver print

71a

CHARLES E. RILLIET *Information booth for Los Angeles International Air Meet*
May 1909
silver print

71b*

CHARLES E. RILLIET *H. Laverne Twining with a modified Wright Brothers glider*
built by students at Polytechnic High School, Los Angeles
1909
silver print

72a

MAURICE BRANGER — *Louis Blériot's equipment arrives at the hangar for Rheims Air Meet*
August 1909
silver print

72b

PHILIPPE MUTIN — *Mrs. Blériot and others watch as Blériot circles the airfield*
August 1909
silver print

72c

MAURICE BRANGER   *Hubert Latham flying at Rheims Air Meet*
in an Antoinette monoplane
1909
silver print

72d*

MAURICE BRANGER   *George Cockburn, the only English entry at*
Rheims, flies by a pylon in a Farman biplane
1909
silver print

ROY KNABENSHUE *Louis Paulhan flying his #2 machine
at Dominguez Hills Air Meet*
January 1910
silver print

75

COLE AND COMPANY   *Arch Hoxsey warping his wings prior to flight*
October 1910
toned silver print

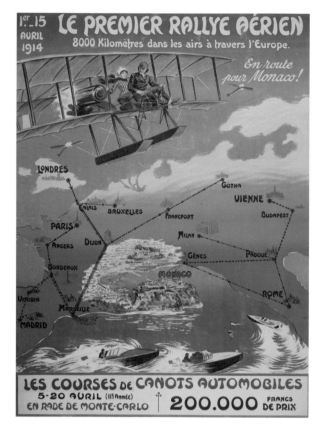

77

COURBET AND COMPANY   *Poster announcing the April 1–15 aero meet*
in Monaco
1914
colored lithograph

78

ERNEST MONTAUT   *Wilbur Wright in flight at Camp d'Auvours,*
Le Mans, France
1908
hand–painted lithograph with letterpress

79

J. NICOL   *Thomas S. Baldwin parachuting from a balloon,*
Monmouth Park, New Jersey
1889
albumen print

80

MRS. JOHN DOUGHTY   *A Balloon in a Storm, published in* Century *magazine*
1886
watercolor and pencil drawing

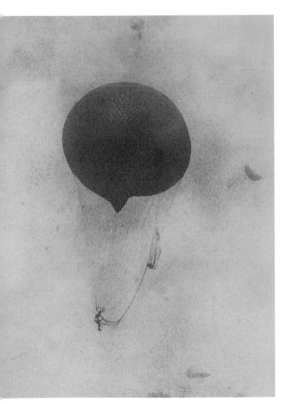

81*

UNKNOWN PHOTOGRAPHER   *Dora Marr and friends at Neuilly-sur-Seine Festival*
posing in a carnival replica of the Spirit of St. Louis
June 1927
silver print

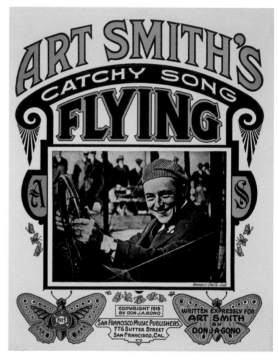

82b

DON J. A. GONO   *Sheet music for "Flying: A Catchy Song*
Written for Art Smith"
1915
photolithograph sheet music cover

82a

UNKNOWN *SAN FRANCISCO*   *Art Smith doing loop-de-loops on the last night*
CHRONICLE PHOTOGRAPHER   *of the Pan-Pacific Expo, San Francisco*
1915
silver print

83a

WILLIAM PRESTON MAYFIELD   *Ruth Law in her Wright flyer*
circa 1912
silver print

*83c**

UNKNOWN AMATEUR PHOTOGRAPHER   *Ruth Law doing loop-de-loops at night,*
Sedalia, Missouri
circa 1917
silver print

83b

UNKNOWN AMATEUR PHOTOGRAPHER   *Ruth Law stunting at Sedalia, Missouri*
circa 1917
silver print

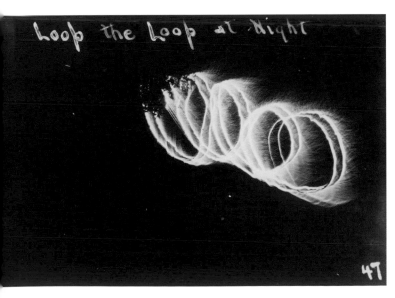

86*

HOLMES *Portrait of Amelia Earhart at the
American Embassy, London*
1928
silver print

85

WILLIAM PRESTON MAYFIELD *Ziegfeld girls lined up next to Roscoe Turner's
Sikorsky plane known as the "Flying Cigar Store"*
circa 1925
silver print

98

HERBERT MATTER *Fashion shoot with airplane*
circa 1946
silver print

88

UNKNOWN PHOTOGRAPHER

A Family Flying Above Ohio
circa 1915
mixed watercolor–and–photo collage

97

UNKNOWN PHOTOGRAPHER

*Soviet propaganda photograph showing fliers
in the sky with a cutout of a twin-engine airplane*
circa 1940s
silver print with overlay

91

ELLIOT SERVICE COMPANY *Experimental pickup of airmail by Adams Air Mail Company*
1930
silver print on board with letterpress

99

HOMCO COMPANY *The Red Baron as depicted in* Peanuts
1975
metal wall sculpture

104

UNKNOWN MAKER *Three-dimensional model biplane*
circa 1920
wire

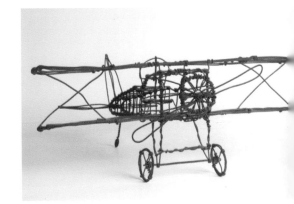

Planes Take Mail Without Stop

Daily airmail service without the necessity of plane landing to take on mail sacks was recently inaugurated at Beaver Falls and New Castle. The Adams Air Mail Pick-up is used and operates as shown in this fine action photo.

WATCH THIS BULLETIN FOR LATEST NEWS

92

UNKNOWN PHOTOGRAPHER   *Woman stunting*
circa 1920s
silver print

93a

WESLEY DAVID WALKER   *World War I hoax, planes approaching crash*
circa 1933
silver print

93b

WESLEY DAVID WALKER   *World War I hoax, planes crashing*
circa 1933
silver print

93c*

WESLEY DAVID WALKER   *World War I hoax, pilot falling from German plane*
circa 1933
silver print

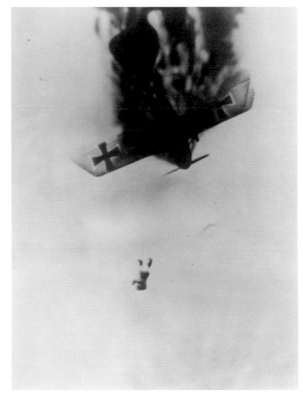

96

UNKNOWN PHOTOGRAPHER · *Carl Groth, age 6, holding a replica of the Graf Zeppelin purchased for him by his uncle in Germany and brought to San Francisco on the original Graf Zeppelin*
1929
silver print

95

UNKNOWN PHOTOGRAPHER · *Roy Knabenshue's airship in flight above the Raymond Hotel in Pasadena, California*
circa 1912
silver print

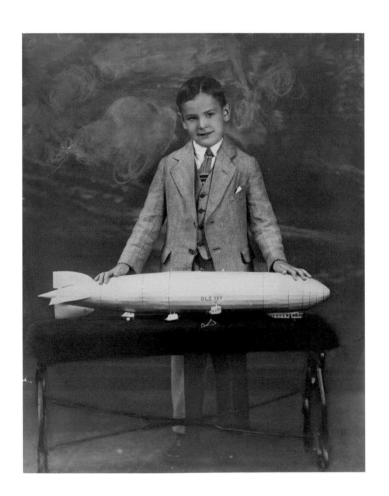

100a–d

UNKNOWN PHOTOGRAPHERS   *Postcards depicting planes as props for tourists*
circa 1915–30
silver prints on postcard stock

FLYING HIGH
AT CEDAR POINT

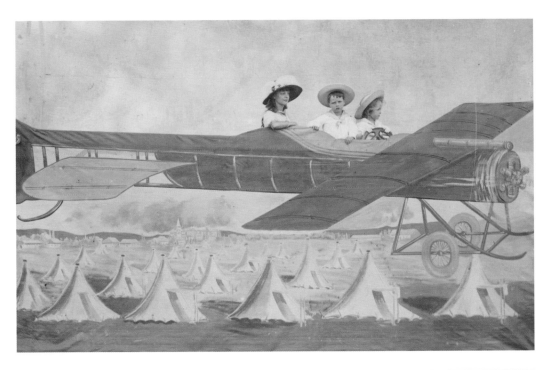

101

UNKNOWN ARTIST *French tapestry depicting Lindbergh and others who attempted to fly the Atlantic*
circa 1920s
fabric

PART II:
THE EARTH FROM ABOVE

42

RISING UP, LOOKING DOWN

STEPHEN WHITE

With the exception of a few odd ducks such as Leonardo da Vinci (who sought to fly like the birds), most of the scientific time and energy spent in the centuries preceding successful flight were dedicated to understanding the moon, stars, and sun and the relationships among them rather than finding a way to fly closer to them. Geometry and astronomy were new fields. Heights were associated with mountains and buildings, whereas distances were measured through the spyglass or the telescope. For Nicolaus Copernicus, Johannes Kepler, Galileo Galilei, Tycho Brahe, and other curious early pioneers of astronomy, the goal was to reach out to examine the universe and then explain what they learned in scientific terms. They did this even at the risk of being put to death as heretics: for them, finding and seeing the true state of things was as important as life.

The bird's-eye view had been used as far back as the fifth century to show perspective in the organization of a city (Jerusalem), and it became a staple of map-making after the discovery of printing in the fifteenth century. Centuries later the bird's-eye view evolved into the aerial view, seen from an altitude even higher than a bird might fly, as illustrated by the images in this section showing the spread of Washington, D.C., from 10,000 feet, or Dayton from 30,000 feet.

In William Langewiesche's book *Inside the Sky: A Meditation on Flight*, he comments on how the view from above the Earth has reshaped our thinking:

> The world beneath our wings has become a human artifact, our most spontaneous and complex creation. Tourists may not like to contemplate the evidence with its hints of greed and self-destruction, but the fact remains that the old sterilized landscapes—like designated outlooks and pretty parks and sculpted gardens—have become obsolete, and that it is largely the airplane that has made them so. The aerial view is something entirely new. We need to admit that it flattens the world and mutes it in a rush of air and engines, and that it suppresses beauty. But it also strips the façades from our constructions, and by raising us above the constraints of the tree line and the highway it imposes a brutal honesty on our perceptions. It lets us see ourselves in context, as creatures struggling through life on the face of a planet, not separate from nature, but its most expressive agents. It lets us see that our struggles form patterns on the land, that these patterns repeat to an extent which before we had not known, and that there is a sense to them.

114a
NASA
Skylab 4 photo of California taken from about 260 miles up
1974
chromogenic print

115b
NASA
TIROS-3 photo from West Cuba to the Yucatán Peninsula
1961
silver print

COPYRIGHT 1906
GEO. R. LAWRENCE CO.
CHICAGO.

RUINS
NOR
FROM LA
180

110

GEORGE R. LAWRENCE
AND COMPANY

*View of San Francisco earthquake destruction
from a captive airship*
1906
photogravure

108*

MICHAUD & CHRÉTIEN,
PHOTOGRAPHERS

L'Assault; troops advancing during the
Battle of the Somme
1916
toned silver print

113

UNDERWOOD AND UNDERWOOD

Atomic bomb exploding in the air over Nagasaki,
taken from about 30,000 feet
1945
silver print

109

AÉRO-PHOTO *Place de l'Etoile, l'Arc de Triomphe, Paris, 1930*
reproduced in André Warnod's *Visages de Paris*
1930
silver print

*107**

JOHN DEERING *Fisheye lens view of New York City from the*
top of the Empire State Building
1953
toned silver print

114b

NASA   **Skylab 4** *photo of Los Angeles vicinity*
1974
chromogenic print

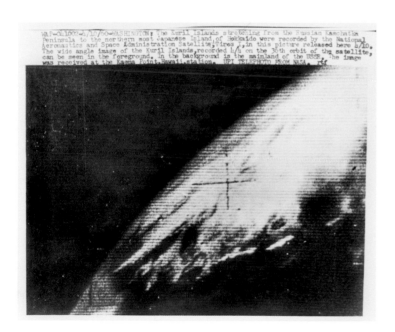

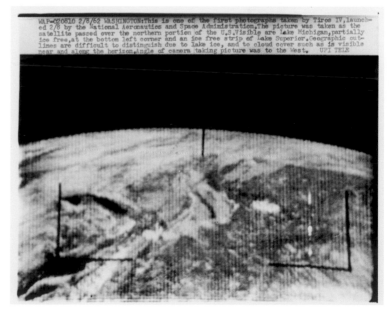

116a

NASA   *Rendezvous of* Gemini 6 *and* Gemini 7
1965
silver print

116b

NASA **Gemini 8** *approaches Agena Target Vehicle*
during historic rendezvous and docking maneuver
approximately 200 miles above Earth
1966
silver print

117

NASA *Earth from sunrise to sunset from a height*
of 22,300 miles above South America,
taken by the ATS–3 satellite
1967
silver print

118

NASA FILE NEGATIVE *Earth, taken from the moon*
1969 (printed by Dennis Ivy circa 1990)
dye transfer print

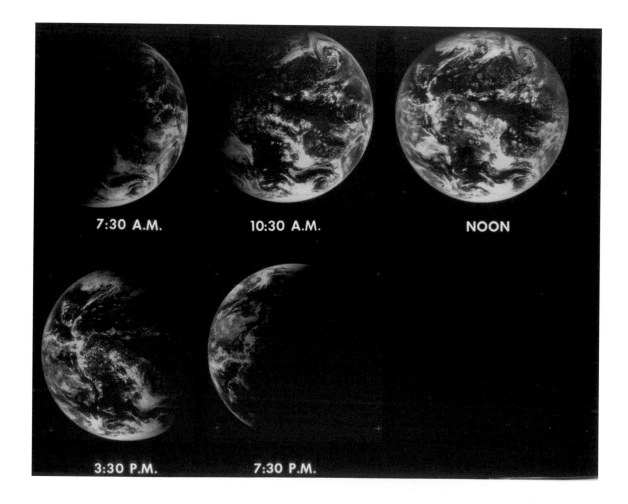

119

NASA   *The Earth seen as a pale blue dot from* Voyager 1,
taken approximately 3.7 billion miles from Earth
1990
digital color print

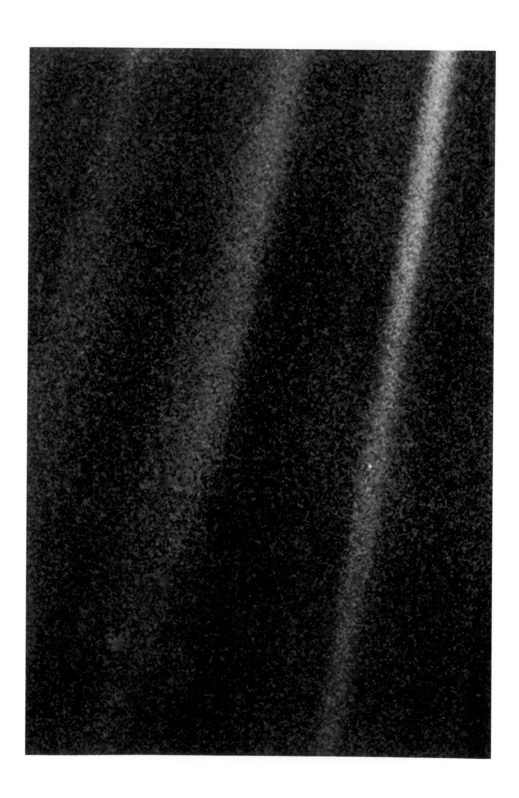

PART III:
SPACE EXPLORATION

LANDSCAPE AND TRAJECTORY

MICHAEL BENSON

Future historians may perceive that the cascade of insights and innovations that finally led to powered flight at the turn of the twentieth century—and ultimately, to space travel—occurred virtually simultaneously with the invention of a means to record it. In fact, by the time the Wright Brothers' flimsy first airplane lifted into the stiff headwinds of the Kitty Hawk peninsula on December 17, 1903, photography was sixty-four years old. The medium's relative sophistication at the time of that first flight is evident in the clarity of our image of the event. Although it was captured on a glass-plate negative, emulsions were sufficiently sensitive and shutter speeds fast enough by then to freeze the Wright Flyer in time, suspended above the sand. The only visible blurring in the image was produced by the craft's two whirling wooden propellers.

The original Flyer can be found at the Smithsonian Institution's National Air and Space Museum. Its delicate spars, cables, and flimsy canvas make it seem more a sketch than a fully realized concept—the skin of an idea rather than its fruit—despite the photographic evidence to the contrary (because, after all, in the image the machine is palpably aloft). This documentation of the Wrights' historic flight is among the best known of the approximate 200 prints presented in *Skydreamers*. In a mounted print evidently once owned by Wright Brothers biographer John R. McMahon and used so frequently for mass media reproduction that it bears the patina of many a newsroom and publisher, John T. Daniels's photograph of the miraculously suspended Flyer hangs immutably within the first third of the catalogue.[1]

Skydreamers traces the history of flight from its earliest manifestations in balloons and dirigibles through gliders, heavier-than-air powered flight, advanced airplanes, and finally space exploration, including robotic exploration documented by images from semi-autonomous, unoccupied spacecraft. It's an extraordinary trajectory.

One could say that both art and its cousin, documentation, are surpluses produced by something that itself constitutes a surplus: life itself, a product of excess energy and fortuitous chemistry within the universe's larger frame. Certain patterns in both surpluses are clear. The upward arc documented by *Skydreamers* illustrates an inexorable bursting-out, a vital push by both life and its means of expression to transcend geography, gravity, and even mortality.

If we disregard various creation mythologies in favor of scientific explanations

1. Daniels, a member of the Kill Devil Hills Life-Saving Station, was no photographer; in fact, this is the only photograph he is known to have taken. Conscious of the necessity of good documentation, Orville Wright had pre-positioned the camera and instructed Daniels to squeeze the bulb, triggering the shutter at the right moment.

(which at least have the humility to admit where they fall short), we can say that life was produced through an obscure alchemy in a region of the universe where conditions were right, excess energy abounded, and serendipity reigned. And after 3.7 billion years of incremental evolution—a period punctuated by several near-apocalyptic natural catastrophes that came close to wiping it out altogether—it eventually produced intelligence, which is another way of saying "matter capable of self-regard." A piece of the universe capable of perceiving itself, in other words, had arisen. According to this view, the self-consciousness of the species is the self-consciousness of the universe itself, and that tendency is both abetted and made manifest by our modes of expression and documentation. Every photograph is an investigation of matter *by* matter.

124
Unknown photographer
Detail study, #7 of ten different phases of the moon on a single board
circa 1900
silver prints

So one could do worse than survey photography for clues as to our progress, progress both in discerning larger universal truths and in achieving insights about our situation and ourselves. At least in its early phases, before techniques of photographic manipulation became sophisticated, photos were understood as confirmation of something thereby verified as true. The clarity of the shot revealing the suspended state of the Wright Flyer proved to the world that an improbable event had taken place. It was an incontrovertible fact, even though only five witnesses were physically present. Photography was, and in many cases still is, taken as evidence of objective truth.[2] It is a manifestation of, and a means to the furtherance of, the scientific revolution that made empiricism virtually the only means for studying natural phenomena. The verification of the visual evidence of the world and of the greater universe inaugurated by photographic processes runs in parallel with and augments photography's impact on the arts and popular culture. By appropriating both the technique (photography and film) and subject matter (flight and space exploration) presented here for their own ends, cultural producers have both benefited from and further propagated interest in things airborne and interplanetary.

Particularly within science and technology, photography can further be understood as a critical element in a process of discovery, reification, and rehearsal. We discern truths about the material world, reaffirm what has been accomplished, and use this as a basis for continuing progress. Nineteenth-century studies of the moon, one of which is included in *Skydreamers*, were critical to establishing the beginnings of a knowledge base later necessary to plan the first remote-controlled robotic landing by a spacecraft on another sphere more than a hundred years later. A grainy two-frame mosaic of a

2. Clearly, digital manipulation has altered immeasurably our understanding of photography in the last decade or so, and prior to that some totalitarian regimes, among other actors, took photographic manipulation to a high art. But certainly with scientific photography (including astronomy) as well as photography conducted in controlled circumstances or with multiple corroborative cameras present, photographic evidence is still considered to be close to incontrovertible.

boulder-strewn landscape rimmed by lunar hills, an image taken in June of 1966, helped inaugurate a new genre: extraterrestrial landscape photography conducted by automated machine.[3] It's a field brought to new levels of refinement over the past six years by NASA's twin Mars Rovers, *Spirit* and *Opportunity*, the first mobile landscape photographers on another world. Like most U.S. robotic spacecraft, the Mars Rovers were designed and built at the Jet Propulsion Laboratory in Pasadena, California. These lunar pictures were in turn themselves necessary in the final push to land human beings there, an event accomplished by *Apollo 11* in 1969.

This process of landscape and trajectory, of photo-augmented research and then further development, is evident throughout *Skydreamers*. The glider experiments of Otto Lilienthal, who made more than 2,000 flights from 1891 until his death in a crash

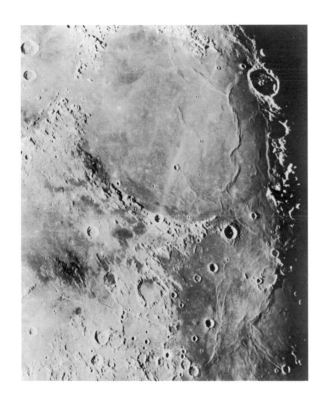

in 1896, are documented by eight shots (a couple of these look remarkably as though they have been transposed directly from the inked lines of Leonardo's notebooks into the silver silt of the photographic medium). These are followed by a pair of prints of similar glider experiments conducted by the Wright Brothers, who credited Lilienthal as a major source of inspiration in their quest for powered flight, and finally, a photograph of the brothers' epochal first flight. The progression here is obvious, and the information transmission necessary for it was at least partially accomplished by the very photographs on view in the catalogue.

Multiple larger and smaller trajectories arc throughout *Skydreamers*. The 1876 photograph of the moon mentioned above links to a remarkably detailed astronomical plate of the moon dating from 1903, that same inaugural year of aviation. In it we see two lunar "seas," Mare Serenitatis and Mare Tranquillitatis, the future locations of three of the six manned landings (*Apollo 11, 15,* and *17*). This in turn finds resonance in an entertaining picture of a coat-and-tie-wearing museum curator literally straddling the plaster-relief mountain range between Mare Serenitatis and Mare Imbrium as he touches up the latter's distinctive Autolycus crater; his shadow extends upwards towards the Lunar North Pole on a vast, semispherical relief map of the Moon's Earth-facing hemisphere. Neil Armstrong, it seems, was the *second* man on the moon.

Completing this particular arc, the catalogue shows three key examples of photographic documentation of *Apollo 11*. They include two Associated Press (AP) wire photos released within five days of the mission's return to Earth on July 24, 1969. These reproduce frames from an automated 16-millimeter motion picture camera that documented

3. These were not the first photographs from the surface of another sphere, however: the Soviet *Luna 9* probe had landed five months previously, on February 3, 1966, accomplishing both the first survivable landing by a spacecraft on the surface of another celestial body and the first extraterrestrial surface photographs taken.

120
Unknown photographer
*Neighbor child looking through a
telescope on the Henry Draper estate,
Hastings-on-Hudson*
1883
albumen print

SPACE FANTASIES

The following quote appears in *The Watershed*, a biography of Johannes Kepler written by Arthur Koestler:

> There will certainly be no lack of human pioneers when we have mastered the art of flight. Who would have thought that navigation across the vast ocean is less dangerous and quieter than in the narrow, threatening gulfs of the Adriatic, or the Baltic, or the British Straits? Let us create vessels and sails adjusted to the heavenly ether, and there will be plenty of people unafraid of the empty wastes. In the meantime, we shall prepare, for the brave sky travelers, maps of the celestial bodies—I shall do it for the moon, you, Galileo, for Jupiter.

Only a year after Galileo invented his telescope, he and Kepler had already begun to dream about what could be discovered through the telescope's lens. Before humans could explore outer space, they needed to locate, chart, and identify. Astronomy became popular in Europe, and by the nineteenth century observatories had sprung up all across the globe. None, however, had telescopes equal to the one seen in the Will Connell photograph depicting the mighty 200-inch lens on the Hale Telescope at Mt. Palomar—until 1993 this was the largest earthbound telescope in the world. By then, the mighty Hubble had been placed in space, ready to study the universe in far greater detail than any earthbound counterpart could. The Hubble would soon open whole new galaxies to our hungry eyes.

121
Charles Marville
*Astronomers with new telescope,
Meudon, France*
1876
albumen print

122

WILL CONNELL   *Astronomer looking through the Hale Telescope*
at Palomar Observatory, California
1949
silver print

123*

UNKNOWN ARTIST   *Imagined lunar point of view of the Earth,*
"Full" Earth and "New" Earth
circa 1860
woodcut

124

UNKNOWN PHOTOGRAPHER   *Ten different phases of the moon on a single board*
circa 1900
silver prints

125

EDWARD L. ALLEN AND FRANK ROWELL   *The moon, made at the Observatório Nacional,
Córdoba, Spain*
1876
carbon print

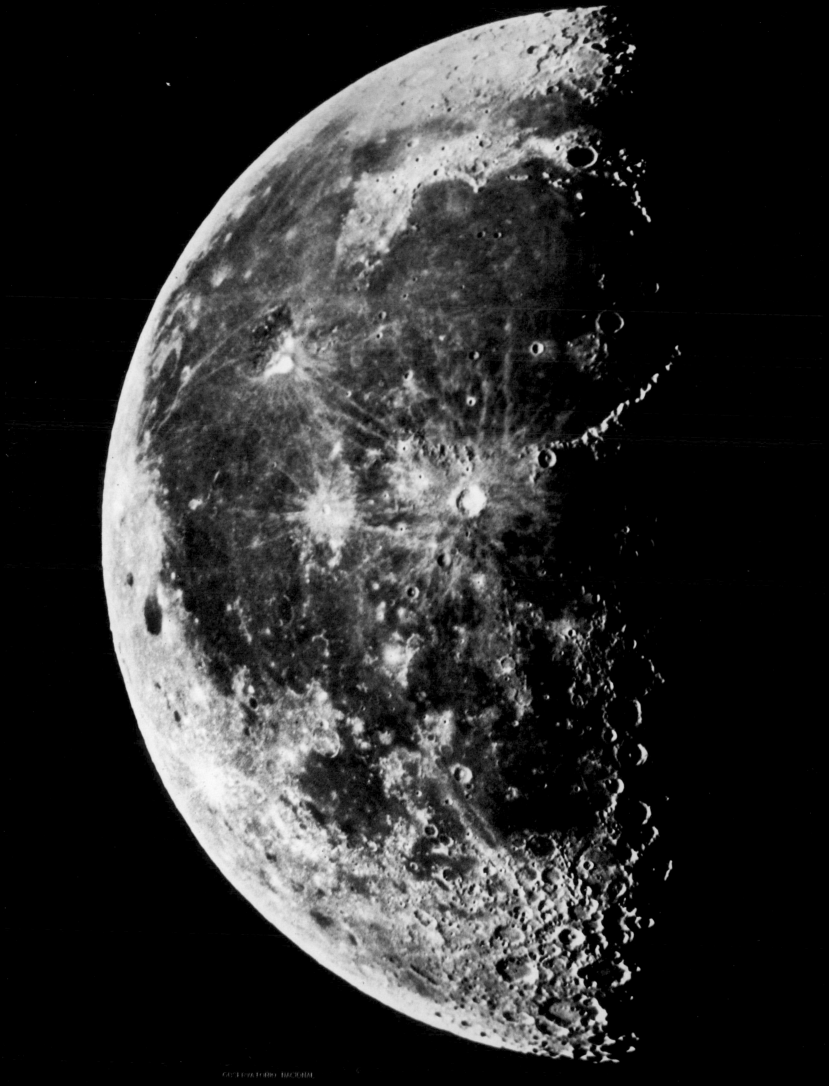

128

FERDINAND QUÉNISSET   *Jupiter and its moons*
circa 1897
gelatin silver print

129

UNKNOWN PHOTOGRAPHER   *The Great Comet (Biscara), seen by the naked eye in the Southern Hemisphere*
1901
gelatin silver print

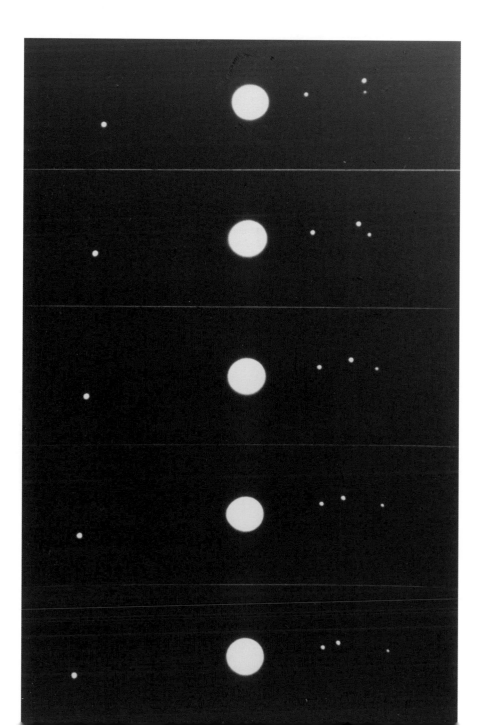

130

UNKNOWN PHOTOGRAPHER *Pleiades, showing nebulous matter attached*
circa 1900
bromide print

135

UNKNOWN PHOTOGRAPHER *Photograph of Orion Nebula*
circa 1955
chromogenic print

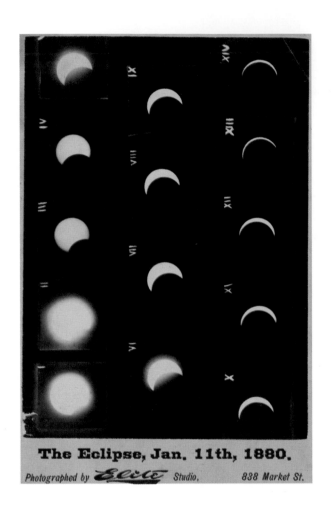

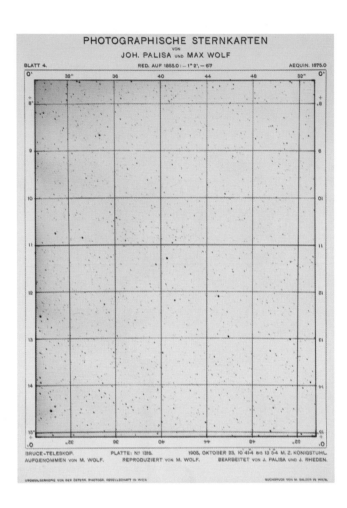

127

ELITE STUDIO, SAN FRANCISCO   *The eclipse, January 11, 1880, showing 14 phases*
1880
albumen prints

133

JOHANN PALISA AND MAXIMILIAN   *Star map plate #1316*
FRANZ JOSEF CORNELIUS WOLF   1905
gelatin silver print

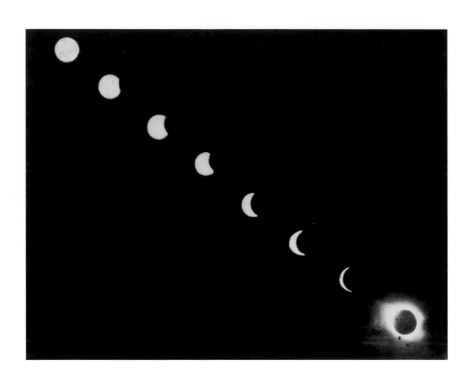

136a

INTERNATIONAL NEWS SERVICE   *A solar eclipse recorded in its stages with a*
special 28–inch lens at Fryeburg, Maine
1932
silver print

136b

INTERNATIONAL NEWS SERVICE   *Full eclipse shot from an airplane at 18,000 feet*
1932
gelatin silver print

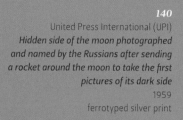

138*
Burt Glinn
*Vanguard TV3 launch
and explosion*
1957
silver print

140
United Press International (UPI)
*Hidden side of the moon photographed
and named by the Russians after sending
a rocket around the moon to take the first
pictures of its dark side*
1959
ferrotyped silver print

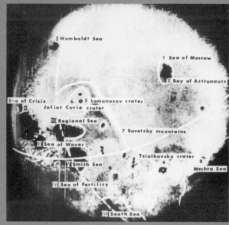

THE SPACE RACE

The space race had two objectives and two principal adversaries: the Russians and the Americans both wanted to be first to fly into space and the first to safely bring an astronaut back from space. In the eighteenth century the Montgolfier brothers sent up animals in their recently invented balloon before they would risk letting a human ride in it. Two hundred years later, the Russians followed suit, initially sending dogs into space. The first dog, Laika, was sent up in *Sputnik II* in 1957, but she was doomed not to return. However, three

years later Belka and Strelka became the first animals to return safely from space. Shortly afterward the Russian Yuri Gagarin was the first man to be launched into orbit. John Glenn, the first American astronaut, was the fifth person in space but the first American to orbit the Earth. In 1962, shortly after Glenn's successful space flight, he testified in Congress against sending women astronauts into space. The Russians apparently disagreed: a year later, the first female astronaut, Valentina Tereshkova, a Russian, orbited the Earth.

139
NASA
*Explorer I satellite on top of
the Jupiter–C launch vehicle*
January 31, 1958
chromogenic print

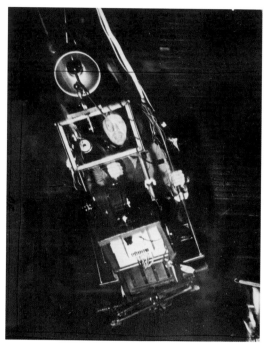

145a

VASILY BATURIN (ATTRIBUTED) *Cameraman in weightless room filming* Star Brothers
circa 1962
ferrotyped silver print

145b

TASS NEWS AGENCY *Testing of centrifuge*
circa 1962
ferrotyped silver print

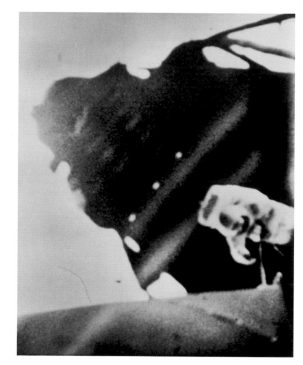

152a

NASA, BUZZ ALDRIN

Neil Armstrong steps down from the lunar module ladder to set foot on the moon
1969
ferrotyped silver print taken from 16mm film

152b

NASA, BUZZ ALDRIN

With the shadow of the module behind him, Armstrong reaches for a camera to record his historic moon walk
1969
ferrotyped silver print taken from 16mm film

153*

NASA, NEIL ARMSTRONG

Buzz Aldrin descending to the surface of the moon
1969
chromogenic print

155
George Cruikshank
*Passing Events, or the Tail
of the Comet of 1853*
1853
etching

SPACE AND POPULAR CULTURE

According to a popular Japanese daily, technology was an answer from God. This Japanese perspective suggested that God had allowed humans to develop technology that would allow them, among other advances, to enter space and move into regions of the universe beyond their own planet. But people had been dreaming of conquering space for centuries, and as we moved closer to achieving that goal during the twentieth century, photography and film informed the world of our progress. As the newest art forms, they triggered the public's fascination and imagination. Not only was Georges Méliès's 1902 film *A Trip to the Moon* the granddaddy of science fiction movies, but its imaginative presentation has stayed in the forefront of the public psyche for more than a hundred years, including in a 1996 music video by the Smashing Pumpkins for their song "Tonight, Tonight." The public reveled in fantasies about space, and the craze in space-related material was spawned by real-life conquests such as the Apollo moon probes and fantasy space adventures like *2001: A Space Odyssey* and *Star Wars*.

159*
Unknown photographer
A flying saucer
circa 1950
silver print

154*

Answer from God, first in a series of articles and illustrations published in the Japanese newspaper Yomiuri Shimbun
1957
silver print with overlays and hand embellishments

UNKNOWN AMATEUR PHOTOGRAPHER   *A group of six photographs taken off the television by a German viewer during* Apollo 11's *trip to the moon and back*
1969
silver prints

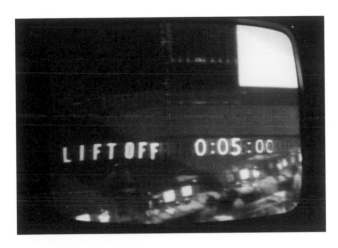

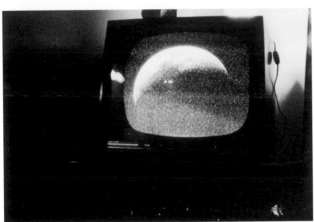

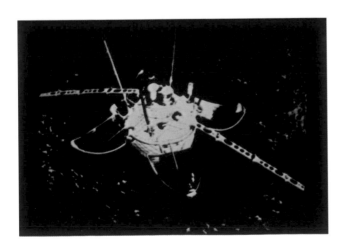

165*
Carlo Naya
St. Mark's Square in moonlight
circa 1870
albumen print

SPACE EXPLORATION AS ART

The examination of outer space was challenging enough to astronomers, and few of them would have considered adding an artistic element to their explorations. Except for the occasional artwork that depicted the moon and moonlight, artists did not begin incorporating elements of the universe into their work until the nineteenth century—only then did stars, planets, and the night sky become popular subjects for artists. Although the best-known night painting is probably Van Gogh's *The Starry Night*, Manet also made an impressive study of the moon over the harbor of Boulogne, and the moon began to be a popular subject for photographers as well. Friedrich Nerly did a painting of St. Mark's Square in moonlight in 1842, and thirty years later Carlo Naya did a series of photographs of moonlight over St. Mark's. Though the moon and stars yielded interesting photographs throughout the twentieth century, it was not until the beginning of the current century that both photographers and painters began stretching artistic ideas about the universe just as they might a canvas. Artists scoured NASA images to find interesting and obscure photographs to print and interpret, and multiple printings led to some of the most interesting contemporary photographs by younger contemporary artists such as the ones in this exhibition: Sharon Harper, Jenny Okun, and Michael Benson.

169
Robert Barrow (attributed)
Moonlight through the moss
circa 1935
toned silver print

170*

LICK OBSERVATORY **Orion Nebula**
circa 1950
(printed by R. P. Peterson)
silver print

172

HELEN LUNDEBERG **Planet,** from the series *The Planets*
1965
oil on canvas

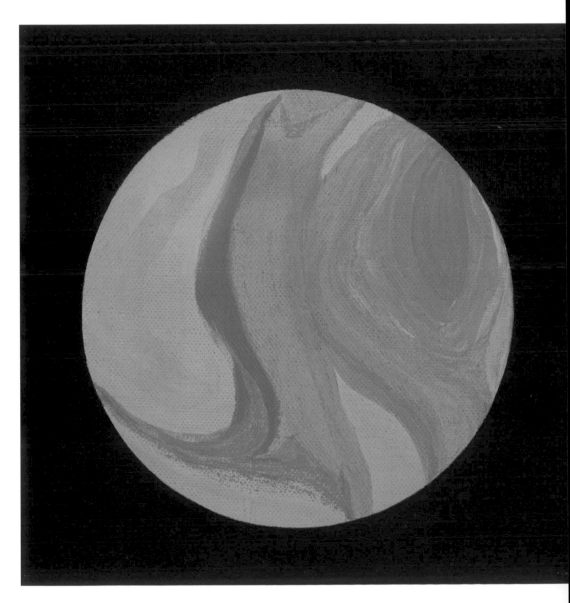

171

FREDERICK S. WIGHT **Moon Descending Beyond Date Palms,**
from the series *Seeing the Light*
1984
oil painting

173

SHARON HARPER **Moon Studies and Star Scratches, No. 9**
June 4–30, 2005, Clearmont, Wyoming;
15–, 30–, 20–, 8–, 5–, 1–, 2–, 1-minute exposures;
15–, 8–, 10–, 14–second exposures;
Luminage print on Fuji Crystal Archive paper, mounted

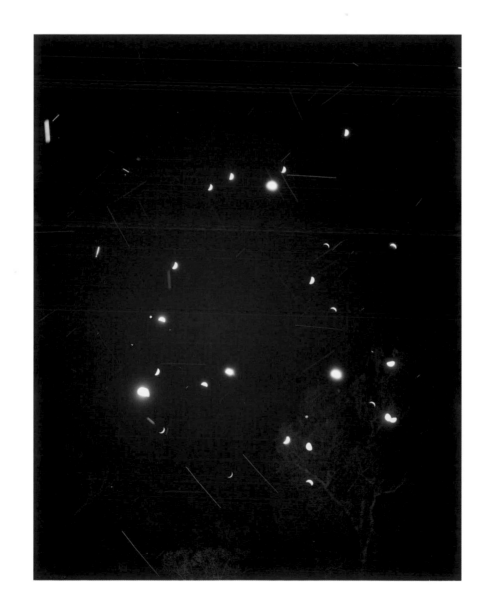

174*

ROBERT WEINGARTEN   *Study of Buzz Aldrin*
2009
digital Epson print

178

DAVID MALIN   *The Sombrero Galaxy*
1990s
chromogenic print

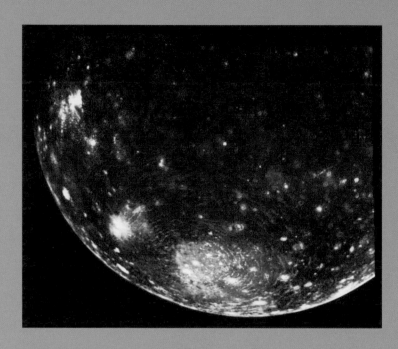

183
NASA
*Mosaic of Jupiter moon Callisto,
taken from* Voyager 1
1979
chromogenic print

PAST THE MOON TOWARD THE END OF THE UNIVERSE

First through the Mariner probes and most recently through images returned by the Hubble Space Telescope, modern astronomers have been able to get an ever greater sense of the scope of the universe. And they have been astonished by what they have encountered: limitless space, endless galaxies, and finally a deep field so far out in time and space that it is almost beyond comprehension. From those first tentative steps into the atmosphere in the basket of a Montgolfier balloon to Hubble's capture of distant spaces, one question lingers: Is it our destiny to one day disappear into nothingness, like a grain of sand flicked into a vast ocean?

180
NASA
*A Phoenix Lander on
the surface of Mars*
1975
chromogenic print

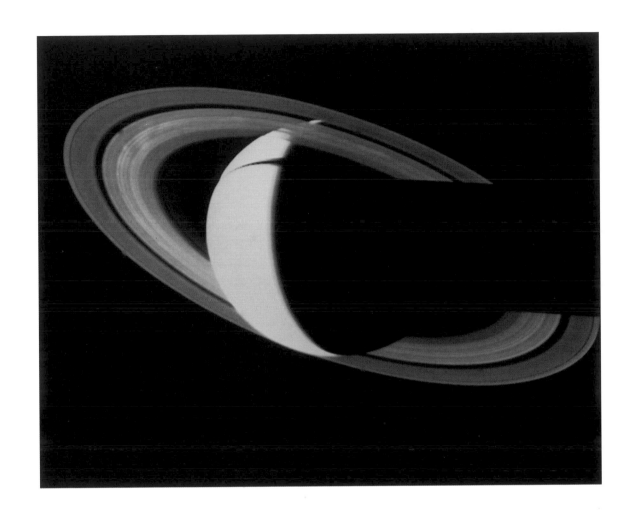

181a

INTERNATIONAL COMMUNICATION
AGENCY

Saturn as seen by Voyager 1 during its approach
1980
chromogenic print

181b

INTERNATIONAL COMMUNICATION
AGENCY

Looking back on Saturn from a distance of 3.3 million miles
1980
chromogenic print

182

NASA   *The Great Red Spot and surrounding area on Jupiter*
1980
chromogenic print

184*

NASA   *Hubble Space Telescope, ultra–deep field,*
constructed from 11 days of observations
2004 (printed 2010)
modern digital color print

skydreamers checklist

FLIGHT FANTASIES

1. Unknown artist
   Expérience de la Machine Aérostatique, 1783
   Hand-colored engraving for *vue d'optique*
   11½ x 16¼ inches

   This first balloon ascent over Paris had no aeronauts aboard.

2. Constance Warburton
   The First Carriage Ariel, 1843
   Watercolor copy of engraving
   8½ x 9¾ inches

3a. Félix Nadar
   Self-portrait in balloon basket, 1863
   Albumen print
   2⅛ x 3½ inches

3b. Félix Nadar
   Le Géant at Invalides, 1863
   Albumen stereo card
   3⅜ x 6⅞ inches

4. Sir Henry James, Ordinance Office
   La Minerve, London, 1864
   Hand-colored photo-zincograph
   11 x 9¼ inches

5a. John Doughty
   Large cumulus cloud, Photography from a Balloon #7, Connecticut, 1885
   Albumen print on boudoir card
   4½ x 7½ inches

5b. John Doughty
   View of the Earth from an altitude of one mile, Photography from a Balloon #5, Connecticut, 1885
   Albumen print on boudoir card
   4½ x 7½ inches

6. Unknown artist
   Balloon Ascension, Japan
   19th century
   Woodcut print
   13¾ x 9¼ inches

7. Unknown photographer
   Balloon ascension, Ferndale, California, 1871
   Albumen print
   5 x 8 inches

8. Unknown photographer
   La France over Paris, seen from Meudon Observatory, 1885
   Photogravure print
   10 x 14 inches

9. Unknown photographer
   Otto Lilienthal wearing his bird wings, 1890s–circa 1910
   Silver print with manuscript letterpress
   12 x 15 inches

10a–g. Unknown photographers
   Otto Lilienthal in flight, circa 1870–90s (printed circa 1920)
   Seven silver prints, each printed on postcard stock
   Each image 2½ x 3½ inches

11. Lorin Wright
   Wilbur Wright gliding, 1902
   Silver print
   6 x 9 inches

   Lorin was the brother of Orville and Wilbur.

12. Rol Agency
   Octave Chanute gliding, circa 1904
   Silver print
   4¼ x 5¼ inches

13. Unknown photographer
   Professor Langley's "Aerodrome" model plane, 1901
   Silver print
   10½ x 13½ inches

   Samuel P. Langley, the director of the Smithsonian Institution, was in competition with the Wright brothers to make the first flight.

14. Harry Ellis
   10 photographs depicting balloon and ascent of Count Charles de Lambert and Mrs. Lambert, circa 1905
   Silver prints
   Each image 4½ x 7 inches

1*
Unknown artist
Expérience de la Machine Aérostatique (detail)
1783
hand-colored engraving
for *vue d'optique*

88. Unknown photographer
A Family Flying Above Ohio,
circa 1915
Mixed watercolor-and-photo
collage
10 x 14 inches

89. Boris Kudoyarov
First airplane in the village, 1930s
Silver print
17 x 23½ inches

90. Unknown photographer
Natives and bomber crew, 1929
Silver print
6½ x 8¾ inches

91. Elliot Service Company
*Experimental pickup of airmail by
Adams Air Mail Company*, 1930
Silver print on board with
letterpress
13 x 16 inches

92. Unknown photographer
Woman stunting, circa 1920s
Silver print
10 x 14 inches
Courtesy of Seaver Center for
Western History Research,
Los Angeles Museum of Natural
History

93a. Wesley David Walker
*World War I hoax, planes
approaching crash*, circa 1933
Silver print
12⅝ x 10 inches

93b. Wesley David Walker
*World War I hoax, planes
crashing*, circa 1933
Silver print
9½ x 12¼ inches

93c. Wesley David Walker
*World War I hoax, pilot falling
from German plane*, circa 1933
Silver print
12½ x 9⅞ inches
All three images courtesy of
Seaver Center for Western History
Research, Los Angeles
Museum of Natural History

Walker published Death in the Air
*anonymously in 1933, claiming
it was the diary and photos of
a deceased World War I fighter
pilot. This hoax, which used
model airplanes to simulate
WWI fighter planes, was not
uncovered until the 1980s.*

94. Roy Knabenshue
*Tourists riding in the gondola of
Knabenshue's dirigible, St. Louis
World's Fair*, 1904
Silver print
11¼ x 9¾ inches
Courtesy of Seaver Center for
Western History Research,
Los Angeles Museum of Natural
History

95. Unknown photographer
*Roy Knabenshue's airship in
flight above the Raymond Hotel
in Pasadena, California*, circa 1912
Silver print
9 x 13 inches

96. Unknown photographer
*Carl Groth, age 6, holding a
replica of the Graf Zeppelin
purchased for him by his uncle
in Germany and brought to San
Francisco on the original Graf
Zeppelin*, 1929
Silver print
9¾ x 7¾ inches

97. Unknown photographer
*Soviet propaganda photograph
showing fliers in the sky with a
cutout of a twin-engine airplane*,
circa 1940s
Silver print with overlay
7⅜ x 9³⁄₁₆ inches

98. Herbert Matter
Fashion shoot with airplane,
circa 1946
Silver print
11 x 10¼ inches

99. Homco Company
*The Red Baron as depicted in
Peanuts*, 1975
Metal wall sculpture
5¼ x 11⅛ inches

100a–d. Unknown photographers
*Postcards depicting planes as
props for tourists*, circa 1915–30
Silver prints on postcard stock
Each 3½ x 5½ inches or 5½ x 3½
inches

101. Unknown artist
*French tapestry depicting
Lindbergh and others who
attempted to fly the Atlantic*,
circa 1920s
Fabric
19½ x 50 inches

102. Unknown artist
*Original piece of stationery
created for Russian artist
Alexander Rodchenko, with an
airplane in the design*, circa 1925

103a–c. *Three original programs:
1909 Los Angeles Air Meet*
8 x 8½ inches
*1910 Dominguez Hills
International Air Meet*
10 x 6¼ inches
1910 Harvard–Boston Aero Meet
9 x 5 inches

104. Unknown maker
*A three-dimensional model
biplane*, circa 1920
Wire
5½ x 13⅛ x 13 inches

PART II: THE EARTH FROM ABOVE

EARTH VIEWS

105. Unknown amateur photographer
*Shot of the Earth and well-
wishers from a balloon over
Cambridge [Nebraska?]*,
circa 1900
Toned silver print
3 x 5½ inches

106. James Black
*Boston as the wild goose sees
it*, 1860
Carbon print
Oval: 7 x 5½ inches

107. John Deering
*View of New York City from the
top of the Empire State Building,
taken with a fisheye lens*,
circa 1953
Silver print
Circular: 13 x 10½ inches

*A fisheye lens is a wide-angle
lens that shows a curved
perspective.*

108. Michaud & Chrétien,
photographers
*L'Assault; troops advancing
during the Battle of the Somme*,
1916
Toned silver print
8¾ x 11 inches

*Taken from about 500 meters
above the battleground, this
image was reproduced in
La Guerre Aérienne Illustrée
magazine on January 3, 1918.*

157. Keystone View Company
Roger Hayward seated on an exact scale of the moon reduced to 38 feet at Griffith Park Observatory, 1934
Silver print
6⅛ x 8⅛ inches

158. NASA, Neil Armstrong
Buzz Aldrin with Armstrong reflected in his face mask and the lunar module behind, developed in 3-D by GAF View-Master, 1969
Chromogenic print
6 x 8½ inches

159. Unknown photographer
A flying saucer, circa 1950
Silver print
8⅞ x 6⅞ inches

Although unidentified, this is most likely a film still rather than the real thing.

162. Jesse F. Santos (attributed)
Star Wars *painting depicting battle,* circa 1980
Original painting on velvet
26 x 54 inches

Santos is an illustrator of comic books.

163. F. G. Weller
What I saw in the Moon, 1876
Albumen stereo card
3⅛ x 6 inches

164. Unknown artist
A disc for a phenakistoscope showing an eclipse, circa 1830
Woodcut
Circular: 7 inches in diameter

A phenakistoscope was an early (1830s) toy that demonstrated persistence of vision.

166. Unknown photographer
The rising and setting sun over Nome, Alaska, possibly on the shortest day of the year, circa 1905
Silver print
7⅜ x 9½ inches

167. Joseph M. Welden
Moon over Fairbanks, Alaska, during a three-hour exposure, 1908
Silver print
6¾ x 9¼ inches

168. D. F. Yerex
Photograph taken during an eclipse over New York, 1925
Toned silver print
11¼ x 19 inches

169. Robert Barrow (attributed)
Moonlight through the moss, circa 1935
Toned silver print
20 x 16 inches

170. Lick Observatory
Orion Nebula, circa 1950
(printed by R. P. Peterson)
Silver print
13¼ x 10¼ inches

Peterson printed extensively for Edward Steichen in the 1950s. Another print of this image was included in Steichen's famous photography exhibition The Family of Man *at the Museum of Modern Art.*

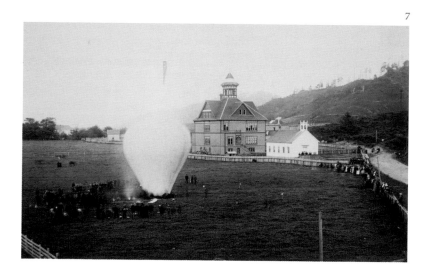

7

160a–f. Unknown amateur photographer
A group of six photographs taken off the television by a German viewer during Apollo 11*'s trip to the moon and back,* 1969
Silver prints
Each 2¾ x 3⅞ inches

161. Unknown photographer
Window card with original poster image from the film 2001: A Space Odyssey, 1968
Photolithograph
16 x 14 inches

SPACE EXPLORATION AS ART

165. Carlo Naya
St. Mark's Square in moonlight, circa 1870
Albumen print
10½ x 13½ inches

Naya created a series of moonlit photographs of Venice.

171. Frederick S. Wight
Moon Descending Beyond Date Palms, from the series *Seeing the Light,* 1984
Oil painting
40 x 60 inches
©Estate of Frederick Wight. Courtesy Louis Stern Fine Arts; reproduced by permission

172. Helen Lundeberg
Planet, from the series *The Planets,* 1965
Oil on canvas
8 x 8 inches
©Feitelson Arts Foundation. Reproduced by permission

173. Sharon Harper
*Moon Studies and Star Scratches,
No. 9,* June 4–30, 2005,
Clearmont, Wyoming
15-, 30-, 20-, 8-, 5-, 1-, 5-,
2-, 1-minute exposures; 15-, 8-,
10-, 14-second exposures
Luminage print on Fuji Crystal
Archive paper, mounted
40 x 50 inches
©Sharon Harper. Courtesy
Galerie Roepke, Cologne,
Germany; Rick Wester Fine Art,
New York; and the artist

174. Robert Weingarten
Study of Buzz Aldrin, 2009
Digital Epson print
40 x 60 inches
©2009 Robert Weingarten.
Courtesy Marlborough Galleries,
New York, and the artist

*From a series of photographs
titled* The Portrait Unbound.

175. NASA, Jenny Okun
From an image series titled Mars
Orbiter, 2006 (printed 2008)
Epson digital paper, composite
print
17 x 40 inches
©Jenny Okun. Courtesy Craig
Krull Gallery and the artist

176. NASA, Michael Benson
*Global dust storm on Mars, taken
from the Viking Orbiter 2, 1977*
(printed 2009)
Digital color print, multiframe
mosaic
53 x 12½ inches
Courtesy NASA, JPL, Dr. Paul
Geissler, and Kinetikon Pictures

177. NASA, Charles Duke, Michael Light
*Image of Charles Duke's family
on lunar surface, photographed
by Charles Duke,* Apollo 16, April
16–27, 1972
Digital C-print; signed, titled,
dated, editioned (edition 50)
24½ x 24½ inches
Transparency NASA; digital image
©1999 Michael Light. Courtesy
Craig Krull Gallery and the artist

178. David Malin
The Sombrero Galaxy, 1990s
Chromogenic print
10½ x 13⅝ inches
©Anglo-Australian Observatory.
Courtesy of Howard Schickler
Fine Art and the artist

PAST THE MOON TOWARD
THE END OF THE UNIVERSE

179a. NASA
*Mariner 6 photo showing a crater
on Mars about 24 miles wide,*
1969
Silver print
6½ x 9¼ inches

179b. NASA
*Mariner 6 photo showing a larger
section of Mars, 430 by 560
miles,* 1969
Silver print
6¾ x 9¼ inches

180. NASA
*A Phoenix Lander on the surface
of Mars,* 1975
Chromogenic print
8 x 10 inches

181a. International Communication
Agency
Saturn as seen by Voyager 1
during its approach, 1980
Chromogenic print
7½ x 9¼ inches

181b. International Communication
Agency
*Looking back on Saturn from
a distance of 3.3 million miles,*
1980
Chromogenic print
7½ x 9¼ inches

182. NASA
*The Great Red Spot and
surrounding area on Jupiter,*
1980
Chromogenic print
15½ x 19½ inches

183. NASA
*Mosaic of Jupiter moon Callisto,
taken from* Voyager 1, 1979
Chromogenic print
15½ x 19½ inches

184. NASA
*Hubble Space Telescope,
ultra-deep field, constructed
from 11 days of observations,*
2004 (printed 2010)
Modern digital color print
20 x 30 inches

*Created over three months, this
photograph shows stars some
13 billion years back in time,
within 400 million years of the
Big Bang.*

acknowledgments

Appropriately, the last page of this catalogue is reserved for acknowledging those who have helped me bring the *Skydreamers* exhibition together. Like the first page of a book or a blank canvas, an exhibition begins with an idea. I had pursued an exhibition on flight for many years before I encountered Jonathan Spaulding, Vice President for Exhibitions, Executive Director, and Chief Curator at the Autry National Center. Jonathan shared my enthusiasm for an exhibition that evolved beyond my own expectations. This project continues the Autry's occasional exploration of the West in global—and now universal—context, and it was Jonathan's vision that made it possible.

Others at the Autry have been instrumental in carrying the idea to its logical conclusion: Andi Alameda, Exhibition Project Manager, who tenaciously made everything stay on course; Stephanie Kowalick, Assistant Registrar, kept track of an ever changing group of images; Scott Frank planned education events; Paula Kessler developed media for the exhibition; Marlene Head spent long hours editing the material for the exhibition and the catalogue; Sandra Odor, Jonathan's executive assistant, connected a lot of the dots; the design group—Julia Latané, Chief Preparator, Mark Lewis, Exhibit Designer, and Patrick Fredrickson, Design Director—worked hard to shape a diverse and extensive checklist into a coherent and compelling composition. Finally, Richard Moll matted and reframed many of the photographs and works of art for the exhibition. My appreciation is extended to them all.

My friend Bill Cartwright was instrumental in encouraging me to continue my pursuit of this idea. Jocelyn Tetel, Vice President in charge of Development at the Skirball Cultural Center, put me in touch with Jonathan Spaulding and others.

While many images came from my own collection, a number of artists graciously lent their work, including Michael Benson, Sharon Harper, Jenny Okun, Bob Seidemann, and Robert Weingarten. Craig Krull of Craig Krull Gallery and Louis Stern of Louis Stern Fine Arts helped facilitate contacts and reproduction rights for the artists they represent.

For inspiration, I point to a brilliant book that fascinated me some years ago, *A Passion for Wings: Aviation and the Western Imagination, 1908–1918* by Robert Wohl, a great read for anyone interested in the broader edges of this subject. And hats off to Corey Keller for her brilliant exhibition (and catalogue) *Brought to Light: Photography and the Invisible, 1840–1900*, shown at the San Francisco Museum of Modern Art and the Albertina in Vienna. Her catalogue was an inspiration for my more modest aspirations.

Finally, thanks to Hugh Milstein at Digital Fusion, Pilar Perez, Michele Perez of Perezdesign, and Mus White for their help in launching this little project into space.

STEPHEN WHITE, SEPTEMBER 2010

174
Robert Weingarten
Study of Buzz Aldrin (detail)
2009, digital Epson print

223